4th International Conference on Mechanical, Manufacturing, Modeling and Mechatronics (IC4M 2019) and 4th International Conference on Design Engineering and Science (ICDES 2019)

IOP Conference Series: Materials Science and Engineering
Volume 520

Nice, France
22 – 24 February 2019

ISBN: 978-1-5108-8553-0
ISSN: 1757-8981

Printed from e-media with permission by:

Curran Associates, Inc.
57 Morehouse Lane
Red Hook, NY 12571

Some format issues inherent in the e-media version may also appear in this print version.

Printed by Curran Associates, Inc. (2019)

For permission requests, please contact the Institute of Physics at the address below.

Institute of Physics
Dirac House, Temple Back
Bristol BS1 6BE UK

Phone: 44 1 17 929 7481
Fax: 44 1 17 920 0979

techtracking@iop.org

Additional copies of this publication are available from:

Curran Associates, Inc.
57 Morehouse Lane
Red Hook, NY 12571 USA
Phone: 845-758-0400
Fax: 845-758-2633
Email: curran@proceedings.com
Web: www.proceedings.com

Table of contents

Volume 520

2019 4th International Conference on Mechanical, Manufacturing, Modeling and Mechatronics (IC4M 2019) & 2019 4th International Conference on Design Engineering and Science (ICDES 2019)

22–24 February 2019, Nice, France

Accepted papers received: 21 March 2019
Published online: 17 April 2019

Preface

Papers

Mechanical, Manufacturing and Mechatronics Engineering

Modeling in Engineering and Design Engineering Science

IOP Conf. Series: Materials Science and Engineering **520** (2019) 011001 doi:10.1088/1757-899X/520/1/011001

PREFACE

2019 4th International Conference on Mechanical, Manufacturing, Modeling and Mechatronics (IC4M 2019) in conjunction with 2019 4th International Conference on Design Engineering and Science (ICDES 2019) were held in Nice, France, from February 22-24, 2019. IC4M 2019 and ICDES 2019 are organized by Hong Kong Society of Mechanical Engineers. The conference provides a useful and wide platform both for display the latest research and for exchange of research results and thoughts in Design Engineering and Science, Mechanical, Manufacturing, Modeling, Mechatronics and other topics. The participants of the conference were from almost every part of the world, with background of either academia or industry, even well-known enterprise. The success and prosperity of the conference is reflected high level of the papers received.

The proceedings are a compilation of the accepted papers and represent an interesting outcome of the conference. This book covers 2 chapters: Mechanical, Manufacturing and Mechatronics Engineering; Modeling in Engineering and Design Engineering Science.

We would like to acknowledge all of those who supported IC4M 2019 and ICDES 2019. Each individual and institutional help were very important for the success of this conference. Especially we would like to thank the organizing committee for their valuable advices in the organization and helpful peer review of the papers.

We sincerely hope that IC4M 2019 and ICDES 2019 will be a forum for excellent discussions that will put forward new ideas and promote collaborative researches. We are sure that the proceedings will serve as an important research source of references and the knowledge, which will lead to not only scientific and engineering progress but also other new products and processes.

IOP Conf. Series: Materials Science and Engineering **520** (2019) 011001 doi:10.1088/1757-899X/520/1/011001

Prof. Garcia Granada Andrés, IQS Barcelona, Spain
Prof. Jan Detand, Ghent University, Belgium

IOP Conf. Series: Materials Science and Engineering **520** (2019) 011002 doi:10.1088/1757-899X/520/1/011002

COMMITTEES

Conference Co-chairs

Prof. Garcia Granada Andrés, IQS Barcelona, Spain

Prof. Du Ruxu, South China University of Technology, China

Prof. Jan Detand, Ghent University, Belgium

Dr. Bertrand.FILLON, the French Plastic and Composites Institute, France

Prof. Cees de Bont, Hong Kong Polytechnic University, Hong Kong

Program Committee

Prof. Yong Zeng, Concordia University, Canada

Prof. Yan Wang, Georgia Institute of Technology, USA

Prof. James Andrew Self, Ulsan National Institute of Science and Technology (UNIST), South Korea

Prof. Anindya Deb, Indian Institute of Science, Bangalore, India

International Technical Committee

Prof. Mohd Hamdi Bin Abd Shukor, University of Malaya, Malaysia

Prof. Yuliang Chen, National Defense University, Taiwan

Prof. Mohd Nasir Tamin, Universiti Teknologi Malaysia, Malaysia

Prof. Sim Kok Swee, Multimedia University, Malaysia

Prof. Amaresh Chakrabarti, Indian Institute of Science, India

Prof. Katz Zvi, University of Pretoria, South Africa

Prof. Zengtao Chen, University of Alberta, Canada

Prof. Kai Cheng, Brunel University, UK

Prof. Ismail Fidan, Tennessee Tech University, USA

Prof. John Gero, George Mason University, USA

Prof. Zaidi bin Mohd Ripin, Universiti Sains Malaysia, Malaysia

Prof. Tetsuo Tomiyama, Cranfield University, UK

IOP Conf. Series: Materials Science and Engineering **520** (2019) 011002 doi:10.1088/1757-899X/520/1/011002

Dr. Tai Kang, Nanyang Technology University, Singapore

Dr. Edzrol Niza Bin Mohamad, University of Malaya, Malaysia

Dr. Mansoor Alghooneh, Immersion Corporation, Canada

Dr. Feng Zhu, Embry-Riddle Aeronautical University, USA

IOP Conf. Series: Materials Science and Engineering **520** (2019) 011003 doi:10.1088/1757-899X/520/1/011003

Peer review statement

All papers published in this volume of *IOP Conference Series: Materials Science and Engineering* have been peer reviewed through processes administered by the proceedings Editors. Reviews were conducted by expert referees to the professional and scientific standards expected of a proceedings journal published by IOP Publishing.

IOP Conf. Series: Materials Science and Engineering **520** (2019) 012001 doi:10.1088/1757-899X/520/1/012001

Effect of Interfacial Adhesion on Tensile Strength of 3D Printed Particulate Nanocomposites

Muhammad Asif[1, a], Maziar Ramezani[1] and Kean Chin Aw[2]

[1]Mechanical Engineering, Auckland University of Technology 55 Wellesley Street East, Auckland 1010, New Zealand
[2]Mechanical Engineering, University of Auckland, Auckland 1010, New Zealand

[a] muhammad.asif@aut.ac.nz

Abstract. Tensile strength of a nanocomposite is dependent on different characteristics such as particle size, particle loading and matrix–particle interfacial adhesion. Matrix–particle interfacial adhesion is one of the main characteristics that affects the mechanical properties. Additive manufacturing have unique and distinct advantages over conventional manufacturing techniques. However, nanocomposites fabricated by additive manufacturing still need a comprehensive study. In this paper we investigated the effect of interfacial adhesion between matrix and filler of samples printed using photopolymer extrusion 3D printing technique. In order to compare the matrix-particle interfacial adhesion of samples printed by this technique, another set of samples have been fabricated using casting technique. Tensile tests on both sets of samples had been conducted and interfacial adhesion parameter models and SEM was used to draw the comparison; consequently, samples fabricated by 3D printing proved to have better mechanical properties than casted samples.

1. Introduction

To overcome the apparent limitations of polymers such as low stiffness and low yield strength, particulate fillers such as micro/nano silica, Al_2O_3, $Mg(OH)_2$, $CaCO_3$ particles, glass and carbon nanotubes are commonly used to manufacture polymer composites. These fillers diversify the mechanical and physical properties of polymers by introducing their rich mechanical behaviour. By adding either micro or nano particles, Young' modulus or stiffness can be dramatically improved [1-3] However strength is heavily dependent on the effective stress transfer between particles and matrix. If the bond between the matrix and particle is good, stress can be effectively transferred from the matrix to the particles. There are several characteristics on which mechanical properties of polymer matrix nanocomposites rely on; i.e. particle size, matrix-particle adhesion, particle distribution and particle loading. Particle loading and matrix–particle interface adhesion affect mechanical properties. For instance, tensile strength of glass bead filled polystyrene composites is dependent on matrix– particle adhesion and increases by improving adhesion [2].

3D printing is an additive manufacturing process in a layer-by-layer additive fashion by either direct material deposition or selective material consolidation within a vat or powder bed. Traditional polymer additive manufacturing techniques such as fused deposition modelling (FDM), fused filament fabrication (FFF) and big area additive manufacturing (BAAM) use thermoplastic material that are liquefied in a heated deposition before being deposited or they used liquid polymers which are cured when exposed to ultraviolet light as in steteolithography (SLA) and polymer jetting. Here, we investigate the effect of interfacial adhesion and distribution of particles on the mechanical properties of 3D printed polymer-matrix nanocomposite. In order to compare mechanical properties of 3D printed polymer nanocomposites, a comparison is drawn by testing specimens fabricated by casing.

2. Background

2.1 Effect of Matrix-Particle Interfacial Adhesion on Tensile Strength

Considering the stress cannot be transferred from matrix to the filler, in this case strength of polymer-matrix nanocomposite is determined by available effective region of load bearing matrix in the absence of filler. Interfacial layer is unable to transfer stress and the strength of nanocomposite is dependent on effective load bearing cross sectional area portion $(1-\varphi)$ as

$$\sigma_c = \sigma_m\,(1-\varphi) \tag{1}$$

where σ_c and σ_m are tensile strength of composite and matrix respectively. Considering φ as a power law function Eq. (1) can be rewritten as:

$$\sigma_c = \sigma_m\,(1-a\varphi^b) \tag{2}$$

where φ is volume fraction and (a) and (b) are constants depending upon particle shape and arrangement in the composite.

Nicolais and Narkis [4, 5] based on Eq. (2) presented a model to predict the tensile strength of nanocomposite reinforced with spherical filler as:

$$\sigma_c = \sigma_m\,(1-a\varphi^{2/3}) \tag{3}$$

When particle and matrix possess a good adhesion, a small portion of the stress is transferred by interfacial region, while the deformation of matrix is very small. In this case, stress is effectively transferred and tensile strength is the combination of matrix and filler properties. Therefore, parameter, *a* in Eq. 3 becomes smaller than 1.21 indicating strong adhesion.

Another model presented by Kunori and Geil [6] associated the tensile strength of composite with parameter , *a* considering it as stress concentration factor as:

$$\sigma_R = exp\,(-a\varphi) \tag{4}$$

where σ_R is relative tensile strength and expressed as $(\sigma_c\,/\sigma_m)$. Higher values of parameter, *a* indicate higher stress concentration.

A model developed by Pukanszky talks about the spontaneous formation of interphase in the nanocomposite considering variation of tensile strength as a function of composition [7]. The model can be written as:

$$\sigma_R = \sigma_m\,\frac{1-\varphi}{1+2.5\varphi}exp\,(B\varphi) \tag{5}$$

Parameter, *B* is expressed as the load carried by the dispersed phase based on its interaction, which can be considered as a measure of matrix-filler adhesion, higher values of B shows good adhesion. The parameter, *B* can be written as:

$$B = (1 + A\rho l)\ln(\frac{\sigma_i}{\sigma_m}) \tag{6}$$

where A is surface area of particles, ρ is the density of particle, l and σ_i are thickness and strength of interphase, respectively.

A model developed by Sato and Furukawa [8] for Young's modulus of nanocomposites containing an adhesion parameter ζ can be written as as:

$$E = \left[\left(1+\frac{0.5\varphi^{\frac{2}{3}}}{1-\varphi^{\frac{2}{3}}}\right)(1-\psi\zeta) - \frac{\varphi^{\frac{2}{3}}\psi\zeta}{(1-\varphi^{\frac{1}{3}})\varphi}\right] \tag{7}$$

$$\psi = \left(\frac{\varphi}{3}\right)\left(\frac{1+\varphi^{1/3}-\varphi^{2/3}}{1-\varphi^{\frac{1}{3}}+\varphi^{2/3}}\right) \tag{8}$$

where ζ of 1 and 0 shows poor adhesion and perfect adhesion respectively

3. Materials and Methods

3.1 Samples Printed using a Photopolymer Extrusion 3D Printing Technique.
A photopolymer extrusion (PPE) 3D printing technique is used to print the samples, Figure 1 shows the PPE printing technique. The entire 3D printing system includes a UV curing system using two UV laser diodes, an extrusion system consists of a peristaltic pump and a nozzle, and a platform with 3 linear motions (x, y and z) and 2 rotary motions (A and B).

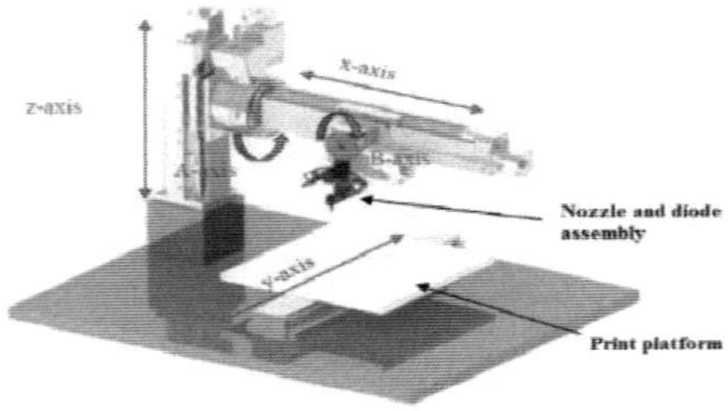

Figure 1: Photopolymer Extrusion 3D Printing Technique

3.2 Samples Fabricated using Casting Technique
In order to compare the mechanical properties of 3D printed samples, samples were prepared using a mould of the type V dog-bone specimen as shown in Figure 2. UV curable resin prepared as discussed in section 3.3 was manually poured into the mould using a syringe and placed under UV light box for 3 minutes. Samples with 8%, 9% and 10% concentrations of the filler were fabricated and tested. Figure below shows the mould used to fabricate cast samples.

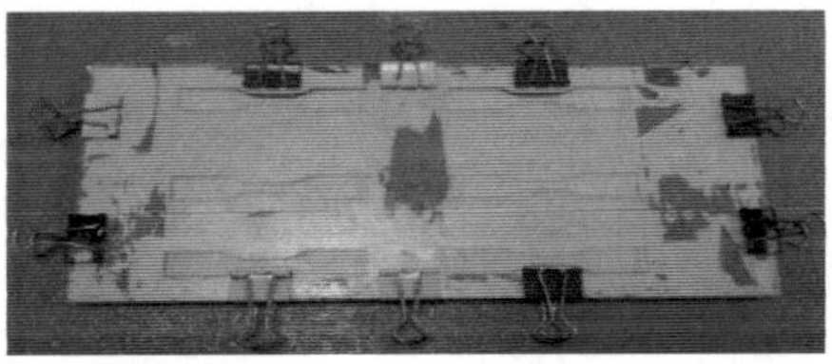

Figure 2. Mould of type V dog-bone specimens.

3.3 Preparation of UV Curable Resin
The photopolymer resin being extruded is made from a commercial polymer named UV Dome 58 (Whitehall Technical Services Ltd) based on an epoxy acrylate that was mixed with fumed silica as filler. In this study fumed silica is used as a reinforcing filler in the resin. For investigation of the mechanical properties of 3D printed samples and samples made by casting, samples were prepared with different concentrations of the silica filler (by weight) mixed into the resin. Mixtures were prepared the same day to have samples with the same aging and were prepared inside a photolithography room. A total of 100 g of mixture was made for each concentration, for instance for 2% by weight of filler, 98 g of the resin was mixed with 2.0 g of fumed silica. After slow manual stirring for 5-10 minutes with a thin spatula, the mixture was mixed with an ultrasonic homogenizer. An ultrasonic homogenizer from Sonics and Materials Inc was used for 2 x 1 minute at 20 kHz of ultrasound frequency and 130 W of intensity. Finally, to reduce air bubbles in the mixture, samples were left to stand in a vacuum oven for 45 minutes at 65 $^{\circ}$C.

4. Experimental Procedure

4.1 Tensile Test

In order to characterize and compare the mechanical properties (e.g. tensile strength and Young's modulus) of 8%, 9%, 10% and pure resin (UV Dome 58) samples, tensile tests were conducted, three dog-bone samples of each concentration of specific dimensions (following ASTM D638 standard type V) were fabricated in accordance with the methodologies discussed in section 3.1 and 3.2. 8%, 9% and 10% filler concentrations were used to print and cast the samples. As the viscosity of the pure resin is too low to be extruded through nozzle, pure resin samples were first cast and then tested. Pure resin has ultimate tensile strength of 7.3 MPa and Young's modulus of 71.6 MPa. Table 1 shows tensile properties of printed and cast samples and Figure 3 shows printed and cast specimen after test.

Table 1. Tensile properties of 8%, 9% and 10% 3D printed samples

Tensile Properties of 3D Printed Samples				**Tensile Properties of Cast Samples**			
Filler Concentration	**Sample**	**Ultimate Tensile Strength (MPa)**	**Young's Modulus (MPa)**	**Filler Concentration**	**Sample**	**Ultimate Tensile Strength (MPa)**	**Young's Modulus (MPa)**
8%	a	32.5	535	8%	a	7.5	195.4
	b	28	422		b	12.0	250
	c	16.7	209		c	13.5	308
9%	a	22.2	507	9%	a	10.3	141.5
	b	34.6	617		b	19.3	256
	c	27.6	623		c	11.8	259
10%	a	28.5	443	10%	a	12.9	231
	b	23.6	386		b	9.9	184.4
	c	25.4	448		c	13.0	249

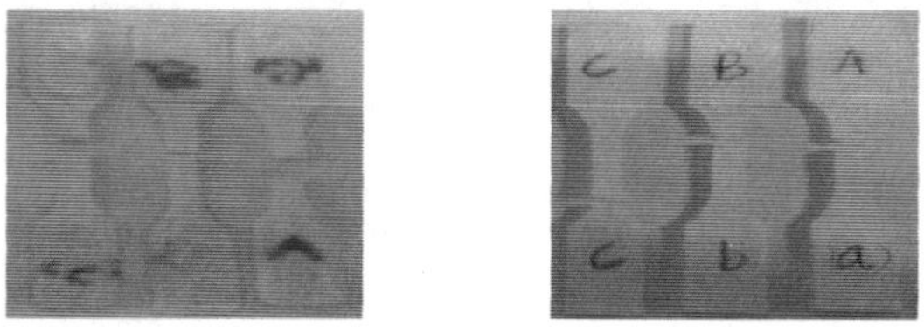

(a) (b)

Figure 3. (a) 3D printed specimen after tensile test (b) Cast sample after tensile test

5 Results and Discussion

5.1 Application of Interfacial Adhesion Parameter Models

3D printed and cast samples were fabricated and tested as discussed in section 3 and 4. Nicolais and Narkis model Eq. (3), Kunori and Giel model Eq. (4), Pukanszky Model Eq. (5) and Sato and Furukawa model Eq. (7) were applied to investigate and compare the interfacial adhesion parameters of 3D printed and cast samples. In order to convert the weight fraction in to volume fraction density of filler (fumed silica) and UV dome 58 (polymer) is taken as 0.18 g/cm^3 and 1.1 g/cm^3 respectively. Volume fractions (φ) 0.347, 0.376 and 0.404 were calculated for 8%, 9% and 10% weight fractions respectively. Figure 4 (a, b & c) represents the boxplots of the calculated interfacial adhesion parameters for 8%, 9% and 10% filler concentrations respectively based on Eq. (3), Eq. (4), Eq. (5) and Eq. (7)

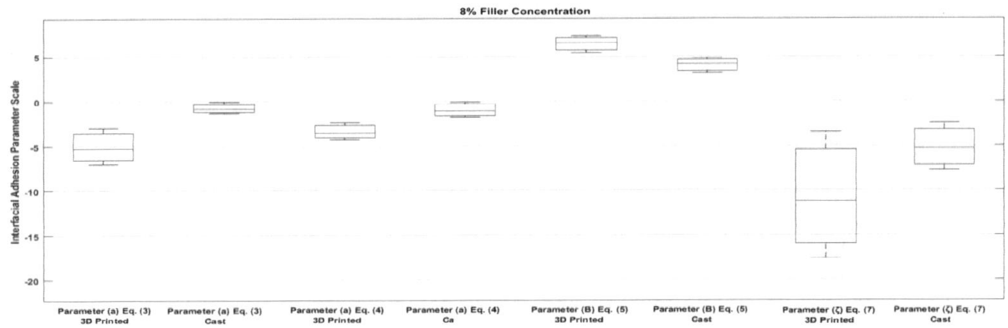

a)

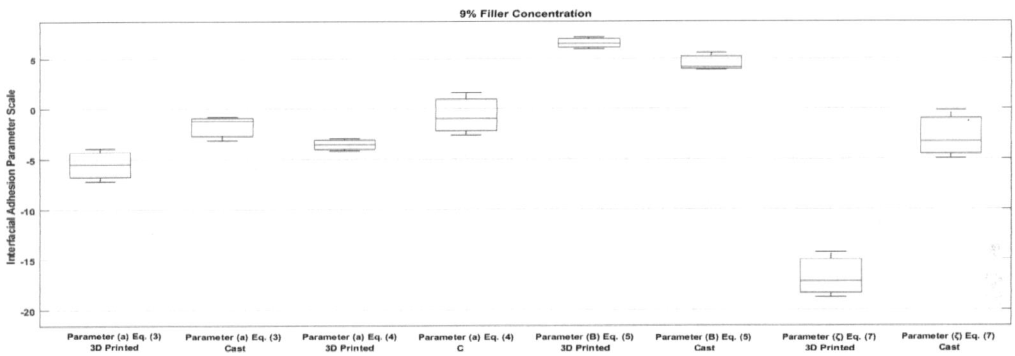

b)

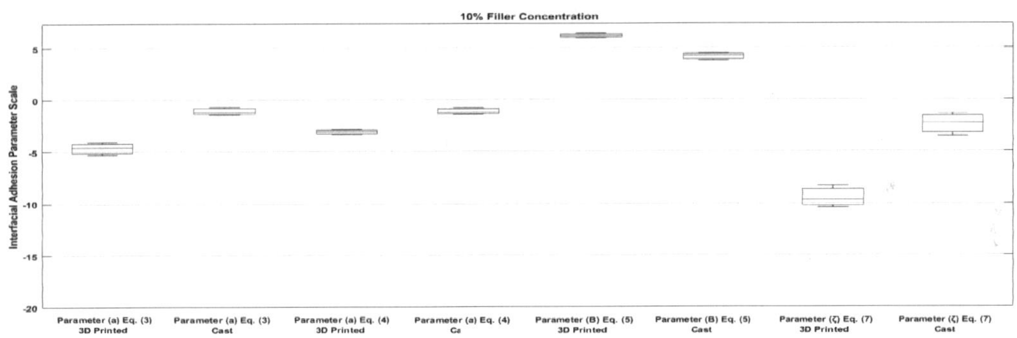

c)

Figure 4: (a) 8% Filler concentration boxplot (b) 9% Filler concentration boxplot (c) 10% Filler concentration boxplot

From Figure 4, it can clearly be seen that all four models predicts very strong interfacial adhesion for 3D printed samples, when seen on the scale defined by each model for perfect and weak adhesion. Cast samples exhibits weak interfacial adhesion as their interfacial adhesion parameters are weaker when compared with 3D printed samples. 3D printed samples for 8%, 9% and 10% filler concentrations exhibits much stronger bond between filler and matrix resulting in higher ultimate tensile strength. In order to further investigate this behaviour of 3D printed samples and to elaborate the reason behind stronger interfacial adhesion, scanning electron microscopy (SEM) was used to study the distribution of the filler at subsurface in both 3D printed and cast samples.

5.2 Scanning Electron Microscopy (SEM)

Figure 5 shows SEM images of the 3D printed and cast samples, it is quite evident from the SEM images that the distribution of the filler in 3D printed samples is more homogenous and the filler is well distributed throughout the image, compared to casted sample where the distribution is not homogenous and clear absence of filler can be seen. This is the reason why the casted samples have weak interfacial adhesion and non homogenous distribution of filler resulted in weak bond between filler and matrix in casted samples. Homogenous distribution of filler in 3D printed samples results in a strong matrix–filler adhesion providing strong bond and ultimately higher interfacial adhesion and mechanical properties.

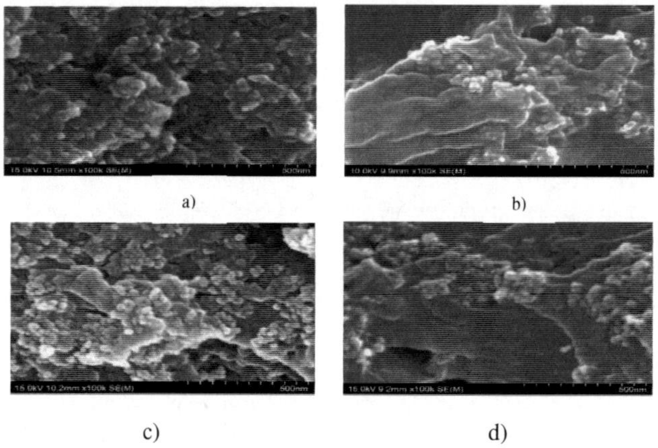

Figure 5: (a & c) SEM image of 3D printed sample (b & d) SEM image of cast samples

6. Conclusion

In this work, interfacial adhesion parameters of 3D printed and casted samples were investigated with the aid of interfacial adhesion parameter models and scanning electron microscopy (SEM). Tensile tests have been conducted and it was found that 3D printed samples exhibit stronger bond between matrix and filler than casted samples as the distribution of the filler is more homogenous in 3D printed samples.

References

[1] Zi-Kang, Z., et al., J. Appl. Poly. Sci., (1999). **73**(14): p. 2977-2984.
[2] J., D.M.E. and H. D., J. Appl. Poly. Sci., (1983). **28**(12): p. 3809-3815.
[3] Fu, S.-Y. and B. Lauke, Comp. Part A: Appl. Sci. and Manuf., (1998). **29**(5): p. 575-583.
[4] Nicolais, L. and L. Nicodemo, Inter. J. Poly. Mate. and Polym. Biomat., (1974). **3**(3): p. 229-243.
[5] E., L.R., N. L., and N. M., J. Appl. Poly. Sci., (1973). **17**(4): p. 1173-1185.
[6] Kunori, T. and P.H. Geil, J. Macrom. Sci., Part B, (1980). **18**(1): p. 135-175.
[7] Pukánszky, B., Comp., (1990). **21**(3): p. 255-262.
[8] Sato, Y. and J. Furukawa, Rubb. Chem. and Tech., (1963). **36**(4): p. 1081-1106.

IOP Conf. Series: Materials Science and Engineering **520** (2019) 012002 doi:10.1088/1757-899X/520/1/012002

Life and Failure of Aircraft Wheels – A Review

Benjamin Rohit[1,*], Punya D Gowda[1], Nandini B Nagaraju[1] and Meghana S[1]

[1]Department of Aerospace Engineering, Rashtreeya Vidyalaya College of Engineering

*e-mail: benjaminrohit25@gmail.com

Abstract. Aircraft wheels mounted with tires support the entire weight of an aircraft during ground operations and landing. Very stringent demands are placed on performance and reliability of landing gear components. The present day wheel is a split type forged aluminium alloy with an internal cold mounted bearing and the brake design energies have increased two fold over the last few decades. The check up for integrity of aircraft wheels are carried out using various techniques by trained and competent personals at regular inspection intervals. From literature we can see that most failures of aircraft rims have been due to high cycle fatigue. Corrosion has played a significant role in fatigue crack nucleation. Braking and high landing loads contribute to the fatigue crack growth. Aircraft's vertical velocity, runway friction coefficient and the asphalts elastic modulus play a significant role in the crack propagation in aircraft wheels. Pitting also serves as a site for stress concentration and hence, nucleation of fatigue cracks. Over-inflation or under-inflation can cause an increase in the stress levels on aircraft wheels.

The present review critically analyses the published work on the fatigue fracture of aircraft wheel rim, current inspection techniques to detect fatigue crack nucleation and growth.

1. Introduction

Wheels and tires of the modern day aircraft are among the most highly stressed parts. Impact loads, cyclic loads, corrosion, and burn outs can contribute to the failure of the landing gear wheels. Failure of wheels could lead to multiple fragments being thrust out, this could pose a great safety threat to maintenance personal or the entire aircraft itself. An aircraft wheel is designed to be able to accommodate the selected tire, to house the brakes and have a long life and minimum weight while accomplishing the above tasks. Manufacturers of this equipment need to adopt a cost conservative design. It has been demonstrated over the past that cost analysis for aircraft wheels and tires needs to be performed based on number of landings than based on the number of flying hours [1].

Forged aluminium alloy is still the most preferred over other alloys of magnesium and steel due to its lower weight and resistance to corrosion.

Titanium alloy wheels can be used if the operator would accept the higher initial cost involved. And with the complexity of forging titanium with the required precision the initial cost could be 10 times higher than that of forged aluminium [1]. Aircraft wheels generally are of two piece type construction to be able to accommodate the very stiff tires that were designed. Wheels need to be tested as per FAA standard for aircraft wheels (TSO-C26c). The wheels are tested to withstand radial loads, axial loads and the combination of both. The wheels also are tested to roll a predefined distance under load and there must be no cracks after the test is completed.

Table 1. A Few Case Histories of Aircraft Wheel Failures

Sl No	Aircraft	Failure Mechanism	Part	Error
1	Lockheed C-130	High cycle fatigue caused the initiation and growth of the crack	Wheel Flange	NDT could not detect crack during initiation [2].
2	Piaggio Avant 180	Corrosion pitting lead to the failure of the inner flange	Wheel Flange	Fretting between tire bead and tube well damaged the protective coating hence, causing corrosion [3].
3	Boeing 737	Fatigue crack initiation at the bore leading to crack growth and catastrophic failure	Inboard Wheel	Stress concentration between the bore and circumferential radius [4].
4	Boeing 737-500	Overheating and fire at the wheel hub caused by the failure of the outboard bearing	Wheel hub and bearing	Incorrect installation of the bearing, insufficient lubrication and the presence of water or contamination could have caused the failure [5].

An FAA study in 1985 concluded that over 30% of 5000 failures reported from 1970 to 1975 were due to wheel problems which did not include the removal of wheels ahead of the service life of the wheel. Fatigue cracks and corrosion are expected as the aircraft progresses through its service life. Corrosion and fatigue cracks could affect the airworthiness of the aircraft. Aircraft wheels are subjected to periodical tests to detect cracks, the most frequently used techniques are non-destructive techniques. Radiography, eddy currents, ultrasound, and penetrants are a few techniques used by very well-trained personal to detect cracks. Eddy current method is the most frequently used method considering the cost and accessibility.

If a crack is detected in a part, it must be repaired or replaced immediately [6]. If a discrepancy is found while the wheel is still mounted on the aircraft, further detailed investigation may be required by removing the wheel from the aircraft.

1.1 Case studies on wheel and tire failures in the past

There have been incidents in the past where the wheel (flange or assembly) or the tire have failed. Wheel failures have occurred due to poor NDT (Non-Destructive Test) techniques, incorrect installation, fatigue and insufficient lubrication or contamination. In many such situations there have been injuries to maintenance and service personal. This section attempts to cover previously failed wheel and tires and the reason to failure. Table 1 gives the summary of case histories of aircraft wheel failures.

2. Aircraft Wheel

Manufacturers of aircraft demand long wheel life capability in terms of roll miles and number of landings/take-offs. Aircraft wheels have two basic configurations, the A-frame and bowl type. Improvement in anti-skid control and greater flexibility in landing gear struts have increased the requirements of the aircraft wheel.

Fatigue failure of the wheel could occur irrespective on the position whether it be the inboard or the outboard half of the wheel assembly. Wheel assembly failures generally occur soon after a tire gets replaced on the wheel and the aircraft is back into service [7].

Aircraft wheels are designed for fail-safe, the point where the tubewell meets the wheel web is generally designed to exhibit low fatigue life and thus behave as a fail-safe point. The fail-safe point is schematically shown in figure 1. A crack which nucleates and grows at this fail-safe point will allow the tire to deflate and this can be detected during a pre-flight inspection [8]. A few reasons for failure of wheel hub are reviewed in this section.

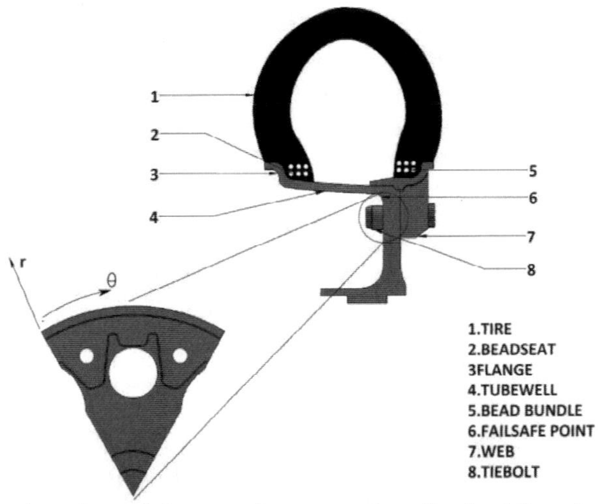

Figure 1. Cross section view of the two piece type aircraft wheel found on modern airliners

2.1 Factors causing crack nucleation in wheel hub

Aircraft wheels are the most susceptible to corrosion than other parts due to the exposure to water and dust. Due to low corrosion fatigue strength of aluminium alloys, corrosion can lead to catastrophic failure. Contact with Airfield pavement de-icing products (PDPs) increase the probability of corrosion in aircraft wheels.

Corrosion pitting is one of the preferred sites for crack nucleation. Corrosion Pitting could also lead to increase in stress concentration and hence acts as a fatigue crack nucleation site. Pitting can be caused as a result of mechanical factors, increased temperatures and surrounding environment [9]. Corrosion pitting could occur even though the wheel is anodized, corrosion could be present before the anodizing process. Corrosion likely forms when the wheel is dipped in a bath during the pre-anodizing process [10].

Corrosion in wheel bearing has been reported in stored aircraft also and the grease used for lubrication played an important role in corrosion protection. It was concluded by an experiment by the US Navy that greases with high alkalinity, good balance of water resistance and water emulsification, resistance to hardening during use and high viscosity greases were good for use of corrosion protection of wheel bearings [11]. The use of the right kind of grease could have prevented the failure of a Boeing 737-500 mentioned in table 1.

Apart from stress concentrations developed from corrosion, stress concentrations could develop at the sharp corners of the fastener hole undercut which is provided for the gear attachment and this stress concentration could lead to a crack nucleation and radial growth of the nucleated crack growth under fatigue loading [12].

Aircraft wheel could also fail due to nut and bolt failure. The nut could fail due to inherent crack or a crack could have applied when it was over torqued during installation or an overload during landing. In some cases the nut gets weakened due to hydrogen effects during the cadmium coating process. Hydrogen embrittlement could occur when cadmium coating is done through electro-deposition plating to protect the component against corrosion. Hydrogen gets attracted to grain boundaries and weakens the grain boundaries. A higher value of stress is required for hydrogen cracking to occur called the static fatigue limit. Failure could when this limit is reached when the nut is either over torqued or when high landing loads are experienced [7]. The lack of cadmium re-plating to the tie bolts could also initiate corrosion fatigue and lead to failure as in the case of a Boeing 737-400 in October 2008[13].Wheel bead seat and flange are also typical regions for crack initiation due high bending stresses and during ground operations stresses peak up as the ground to tire loads are transmitted to the wheel through the wheel tire interface this could provide enough driving force to propagate the crack [10]. When tire seat is fitted against the rim this prohibits the crack from leaking the pressurized gas and thus the crack may grow to critical lengths without being detected. Due to the nonlinear behaviour of the tire, it is also not easy to determine the pressure distribution on the wheel.

2.2 Crack Growth

The nucleated crack could grow as a result of repetitive loading during ground manoeuvres. Braking shortly after landing is one of favourable reason for crack growth. Landing impact stresses also play a significant role in crack growth. In spite of aircraft operators increasing the frequency of inspection to detect defects early catastrophic failures have occurred as the fatigue cracks could not be detected in the early stage [14]. As the cyclic loads on the wheels are applied during ground operations the rate of crack growth is sensitive to the take-off weight of the aircraft. Tire pressure has a strong influence on endurance of the wheel, higher inflation pressure can reduce the life of the wheel by 75% [14].

2.3 Inspection of Wheels during service

All aircraft wheels are inspected for manufacture defects to ensure defect free product before supply. Inspection during the service life of the aircraft wheel is also important. Inspection of the wheel is necessary to detect cracks that could lead to catastrophic failure. Crack detection by NDT is influenced by a few factors such as NDT method, material, configuration, inspector's proficiency and environment.

Location of the cracks plays a vital role in detection by NDT techniques during routine inspection. Locations where the inspection can be conducted without removal of the tire can be conducted as a routine inspection, but few locations such as the tubewell bead seat, flange locations etc. cannot be inspected without the removal of the tire. Cracks developing at the tubewell could be detected during the pre-flight inspection, as a crack in the tubewell will deflate the tire and this can be visually inspected during pre-flight inspection. Crack which have developed in the bead seat and flange regions can only be detected during a tire change.

2.3.1 NDT techniques A few NDT techniques used to detect crack are reviewed in this section. NDT techniques such as the dye penetrant test, eddy current, Ultrasonic and radiography are the commonly used techniques to detect crack during inspection of the wheel hub. It is observed that NDT techniques are highly influenced by configuration of the part, corrosion material, crack location, orientation and tightness. Though the initial investment is high for ultrasonic and eddy current equipment, subsequent inspections are economical. Dye penetrant test is considered to be the most economical of all the NDT tests. But eddy current test is the most widely used technique to detect cracks in aircraft wheels.

Dye penetrant, eddy current and ultrasonic test techniques are not preferable to detect crack with radial lengths less than 2.5mm [15]. Eddy currents are preferable to detect tight cracks where visual, ultrasonic and dye penetrant may fail. Eddy current tests are done around the bead seat area, bolt holes, and the mating surfaces of wheel hubs [16].

Magnetic particle test can only be done on ferrous materials. Tie bolts are tested for cracks by the method of magnetic particle test. Ultrasonic tests can detect cracks through the thickness and can extent through the shot peening region into the susceptible region.

2.4 Aircraft Wheels

While striving to develop the wheel to meet new requirements the cost effectiveness also needs to be considered along with safety requirements. Being very conserved while designing the wheel could add to the cost and also the weight of the wheel. It is a continuous challenge to produce the lightest possible wheel which would meet all the safety requirements at the least cost. Two most important factors to minimize wheel failures are design, test procedures to represent the actual operating conditions and inspection of the wheels. There could be an increase in 10 to 15% of the cost if a certified wheel were to develop a structural problem. Redesigning the wheel, changing forging and retest could be factors that could add to the increase in cost [15]. Additional to the cost due to the change in design, a failed wheel could damage tires and aircraft parts and add to the cost.

The primary wheel cost is generally due to the depletion in fatigue life of the wheel and the timely retirement of the wheel. By controlling corrosion and timely replacement of worn parts can reduce the cost by extending the useful life of aircraft wheels. Careful measures in handling the wheel while storing and transportation could also add to the reduction in cost.

2.4.1 Inspection Cost NDT techniques are commonly used to inspect wheels of aircraft for corrosion and cracks and also for keeping the inspection costs low. A complete check of an aircraft can be a very costly and also takes over a month, and such tests cannot be performed on very short intervals as done on the aircraft wheels.

3. Summary

1. Corrosion, Nut over-torqueing, Hydrogen embrittlement and bolt failures lead to reduction in life of aircraft wheels, and corrosion plays a major role in the fatigue life reduction of aircraft wheels.
2. Corrosion on aircraft wheels is caused due to pavement deicing products (PDP), in- creased temperatures in tires and during braking, pre-anodizing process, quality of grease used etc. Using the prescribed methods aircraft wheels can be protected from corrosion
3. Braking force, landing impact and tire inflation pressure can affect the crack growth rate in aircraft wheels.
4. Eddy current tests, Dye penetrant and Ultrasonic tests are the most used NDT techniques to check for cracks on aircraft wheels, where dye penetrant test is the most economical and eddy current the most reliable.

References

[1] G.E.Stanton, *Journal of Aircraft* **5(1)**, 73 (1968).
[2] T.L. Panontin, *Handbook of Case Histories in Failure Analysis,* Volume **1** (ASM International, 1992), pp. 25–29.
[3] F. Bagnoli, M. Bernabei, *Engineering Failure Analysis* **15**, 654 (2007).
[4] A.T.S. Bureau, ATSB Transport Safety Report AO-2014-109, *Aviation Occurrence Investigation*, Sydney Airport N.S.W (10 June 2015).
[5] I.A. Gaveika, Tech. rep., Transport Accident and Incident Investigation Bureau (TAIIB), Republic of Latvia (18 December 2009).
[6] G. Kosec, G. Kova, J. Hodoli, B. Kosec, Metalurgija **49(4)**, 357 (2010).
[7] T. Kobayashi, D.A. Shockey, *Engineering Failure Analysis* **13**, 65 (2006).
[8] J.A. Sherwood, B.K. Fussell, W.R. Edwards, T.S. Gross, D.W. Watt, *Journal of Aircraft,***32(5)**, 921 (1995).
[9] B. Kosec, G. Kovacicb, L. Koseca, *Engineering Failure Analysis* **9**, 603 (2002).
[10] C. Smith, Air Vehicles Division - DSTO (Defence Science and Technology Organisation) DSTO-TR-**2729** (2012).
[11] A.G. Thomson, Lubrication (1955).
[12] R.Rangaraju, *Metals Materials And Processes*, **12(2)**, 241 (2000).
[13] NTSC, KNKT 08.10.21.04 Boeing 737-400, PK-MDO, South Sulawesi, Republic of Indonesia (2008).
[14] W. Wiebe, National Aeronautical Establishment Ottawa (Ontario) **AD0715751** (1970).
[15] P. Durup, T. Brussat, L. Bakow, Final Report AD-A200 560, FAA Technical Center Atlantic City International Airport N.J. 08405 (1988).
[16] J.K. Jose, *International Advanced Research Journal in Science*, Engineering and Technology **3(5)**, 17 (2016).

IOP Conf. Series: Materials Science and Engineering **520** (2019) 012003 doi:10.1088/1757-899X/520/1/012003

A Review on Inspection and Maintenance of FRP Structures

Rakshith Vishwanath[1] and Benjamin Rohit[1,*]

[1]Department of Aerospace Engineering, Rashtreeya Vidyalaya College of Engineering, India

*e-mail: benjaminrohit25@gmail.com

Abstract. FRP is extensively used today in all domains like concrete structures in Civil engineering, spacecraft in Aerospace Industry. It is also used in elevator and bridge cables. Although FRP has some remarkable properties, it undergoes fatigue and damage and sometimes, undetectable defects are formed. Defects like de-lamination or voids occur due to manufacturing errors, or exposure to external environmental factors can result in BVID. Thus, FRP structures have to undergo inspection to ensure its continued use and safety. The inspection and maintenance methods most commonly employed in FRP structures are NDT or Non-Destructive Testing and CM or Condition Monitoring. The integral part of both, NDT and CM is SHM. NDT is a type of inspection method wherein the sample is tested without damaging it whatsoever. NDT is commonly used for reasons such as low cost, high reliability. The techniques involving NDT are Acoustic Emission Testing, Thermography, Ultrasonic Testing, Eddy-Current testing, Radiography. This paper discusses and reviews the various types of Inspection Techniques in NDT and SHM.

1. Introduction and Application of FRP

FRPs have been extensively used since 1500 BC by the early Egyptians and Mesopotamians and through the variation of properties of FRP, newer products and markets have emerged [1]. The most commonly used FRPs today are GFRP (Glass Fibre), CFRP (Carbon Fibre) and AFRP (Aramids). The domain of application of these fibres is based solely on the properties exhibited by them. In the Construction Industry, FRPs are used widely as reinforcements for concrete structures primarily due to reasons such as high strength, chemical durability, and high corrosion resistance. Carbon Fibres and Aramids are most commonly used as reinforcements [2]. FRPs are also used significantly in the Aerospace Industry in aircraft fuselage and control surfaces and especially as Heat Shields in rockets where CFRP is often used for its high heat resistance [3]. They can also be used in bridge cables and elevator lines for its high strength to weight ratio, good tensile characteristics [4].

2. Defects in FRP and their cause

2.1 BVID

Although FRP exhibits various unique properties in different forms, they are all susceptible to damage and fatigue over time and defects such as BVID or Barely Visible Impact Damage can occur in the structures especially in CFRP [5]. BVID is formed due to Low Velocity Impact which causes delamination internally in the structure and may not be noticeable. These delamination regions grow over time resulting in decrease in stiffness, strength and eventually failure [6]. Failure in a material can either occur during service conditions [7] of the structure or debris impact during runway operations [5] or due to manufacturing errors such as dropped tools [5]. Failure is independent of the fibre properties and crack propagation occurs between the matrix and the fibre due to tension, shear or compressive loading.

2.2 Environmental Factors
Failure due to fatigue occurs due to long time exposure to external environmental conditions such as sunlight, moisture etc. In Paul Böer, Lisa Holliday, Thomas H.-K. Kang [8], CFRP was tested for shear and was immersed in water at different temperatures and significant deterioration was observed upon increasing moisture and temperature and further increase in temperature resulted in greater deterioration. Similarly, in Paul Böer, Lisa Holliday, Thomas H.-K. Kang [8], two different CFRP sheets, one surfaced with a protective coating and the other by epoxy matrix were exposed to sunlight over a duration of 10 years, placed at different sites namely Japan and Canada. The first type of specimen gathered an uneven surface but no internal damage and the second specimen underwent drastic colour change from blue to yellow due to degradation of resin matrix which lowered the bending strength. Exposure to sunlight weakened the intra-ply region between the fibre and matrix. This has been illustrated in Figure1.

Constant inspection, testing and maintenance are very much required in all fields to ensure safe use and service of the structure. Quality testing and regular inspection is the norm to ensure safer and guaranteed functioning of the structures for its stakeholders. The main issue faced is Destructive testing which results in large costs in specimen and test setup. To compensate for the above disadvantage, NDT (Non-Destructive Testing) is used which will be explained in the next section.

3. Introduction to NDT and CM, SHM
Non-Destructive Testing is the testing and inspection of various specimens, materials wherein the material does undergo any damage whatsoever. NDE emphasizes on a range of techniques such as Ultrasonic Testing, Thermography, etc. Almost everything from spacecraft to elevator cables to power stations are inspected using NDT [9]. The overall maintenance of a structure consists of NDT, CM (Condition Monitoring) and SHM (Structural Health Monitoring). NDT is employed during the manufacturing phase of a structure wherein potential flaws can be corrected whereas CM is an ongoing assessment wherein the structure is tested during its operation. SHM is the main objective of maintenance which gathers the results from NDT and CM for the overall maintenance [10].

Relation between NDT, CM and SHM have been shown illustratively in Figure2. NDT has replaced Destructive Testing for its cost, reliability and safety [11].

4. NDT, CM AND SHM Techniques

4.1 Pulsed Eddy Current Measurement
In Yunze He, Gui Yun Tian, Mengchun Pan, Dixiang Chen, [12], PEC or Pulsed Eddy Current has a greater scope and advantage over simple Eddy Current Methods. The specimen used is a Sandwiched Honeycomb Structure. The fibre used is CFRP for its high heat resistance and strength to weight ratio. It is also known for its electromagnetic induction properties hence

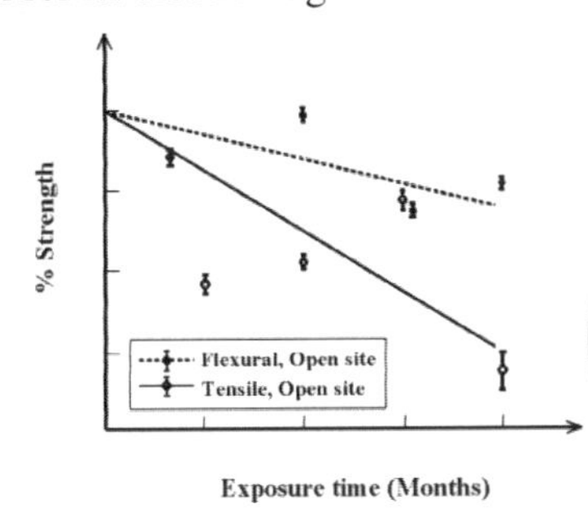

Figure 1. Graph plot of percentage strength and exposure time for FRP

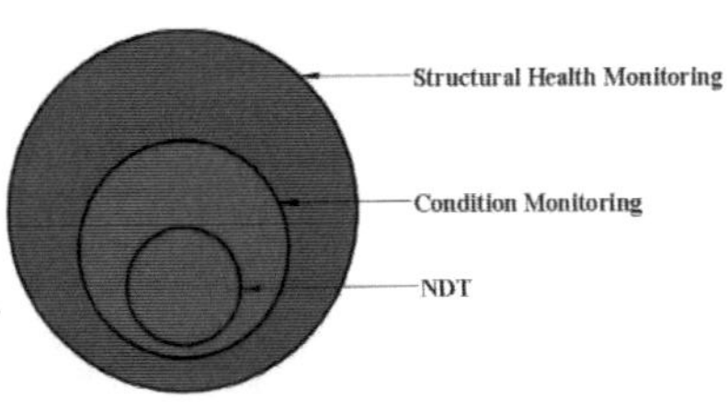

Figure 2. Relation between NDT, CM and SHM

they can be tested by this method [13]. The Eddy Current setup consists of a coil excited by an AC source which is brought towards a sample and hence Eddy Current is induced through Electromagnetic Coupling. The magnetic field created in the sample induces a secondary field in the coil. [14] The defects present change the Eddy Current circulation around the sample and hence changes in magnetic field can be recorded by introducing a secondary coil [13] and thus change in frequency signals are measured from which defects are detected. Figure 3 illustrates different waveforms of different ECT techniques.

4.1.1 Advantages, Disadvantages and Inspection Factors The process is less hazardous, simple, reliable in Surface and Subsurface damage detection, and is economical [12] and in PEC, time required is less [14]. But EDT is very time consuming and cannot be applied to non-conducting materials [15]. Damages such as matrix cracking is undetectable [13]. Factors for detecting defects are defect properties, electrical conductivity of the material. Highly sensitive Magnetic sensors and user-friendly systems can reduce human efforts and time [14].

4.2 Acoustic Emission Testing

In Acoustic Emission Testing, defects can be detected through acoustic emissions generated by a sample on loading. A sample is deformed by loading it and strain energy per unit volume is emitted in the form of elastic waves [16], sensors located on the surface of the sample record transient waves due to deformation [17] such as fibre break, plastic deformation, and crack propagation [18].

Some of the applications of AET are detection of Creep in High Energy Piping (HEP) systems and pressure vessel inspection [19]. Compared to the other general techniques used in NDT, Acoustic Emission Test differs by 2 ways [18], one the signals detected are emitted by the specimen, not by external sources, hence AE test is performed on structures during operation when defect propagation occurs [20], this has been illustrated in Figure 4. Second, strain is measured here, whereas geometrical defects are measured generally in NDT.

4.2.1 Advantages, Disadvantages and Inspection factors Various types of defects, failure mode [18] can be detected by several emission signals and simple equipment is used here [15]. This process involving volumetric inspection is fast [20]. However, distinguishing the various acoustic emission signals produced that overlap over one another is a challenge [21]. Only localized results such as local stress can be detected [15]. The Automation process is complex as it involves linking the transducers and the sample with a special type of fluid [15]. If loading on the structure is not high enough to produce Elastic waves, flaws can sometimes go undetected [20]. Another drawback is high sound generation that affects the signals measured [20].

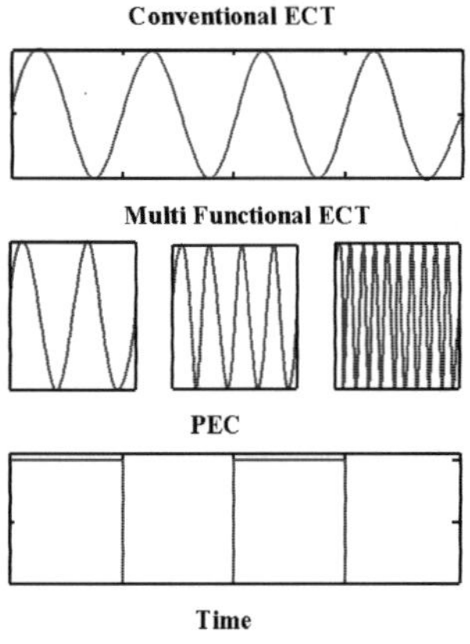

Figure 3. Illustration of excitation waveforms for different ECT techniques

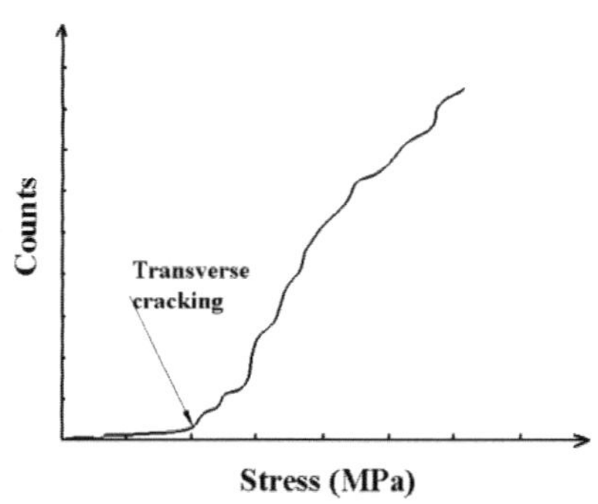

Figure 4. Cumulative AE counts as function of the applied stress

4.3 Thermography

Thermography is the mapping of isotherms over a structure [22] or is the thermal imaging of a structure. In Thermography, defects caused within a structure brings changes in the heat flow which changes the surface temperature at local points [22] and radiation produced is detected by IR cameras.

Detecting flaws using this technique usually depends on detection of thermal gradients in the structure, emissivity of the structure to produce radiation [23]. Thermography is classified as Passive and Active Thermography and both classifications involve use of IR cameras.

In Passive Thermography, the IR camera is placed in front of the sample and a thermal map is generated and in Active Thermography, the surface of the sample is first heated by an external source and the IR camera then records the temperature decay rate as penetration of heat occurs. Defects in the structure show variation in decay rate and surface temperature distribution [22]. Depending upon the decay rate and heat penetration, Active Thermography can be classified as Modulated Thermography (MT), Pulse Thermography (PT) and Pulse Phase Thermography (PPT) [24].

4.3.1 Advantages, Disadvantages and Inspection Factors IRT can scan large surface areas in less time with a single shot [25] and both outer and inner defects can be obtained [15]. The process is simple and a non-contact one [26] and the price of IR cameras used is reasonable today [22]. But, the sensitiveness of inspection is a challenge as some materials have emissivity that results in either rapid heat production or sometimes, undetected defects [13].

4.4 Ultrasonic testing

In Ultrasonic Testing, high frequency soundwaves are propagated towards a sample possessing some acoustic impedance and defects present in the sample reflect the soundwaves back. These sound waves are then displayed on a screen. High frequency waves are used to obtain defects precisely and low frequency soundwaves are used to detect internal defects [27]. The UT setup consists of a receiver which generates high voltage electrical signals [28], an ultrasonic transducer or a probe which converts electrical impulses into sound waves. These sound waves are then propagated towards the sample and defects in the sample reflect the sound waves back to the source which can be visualized on a CRT screen.

4.4.1 Advantages and Disadvantages Results obtained in this method are accurate [15]. Penetration depth for flaw detection is high and equipment used provides instantaneous results [28]. But the entire structure cannot be analysed and experimental setup is costly [15]. Materials with irregular shape, rough surface cannot be tested and a special fluid is required for sound wave propagation [28].

4.5 Radiography

In Radiographic Testing, a sample is subjected to a certain penetrating radiation depending on the nature of the material and a recording medium such as screen projects the defect. For less dense materials such as Aluminium, X ray Radiation is projected and for denser materials, Gamma rays are projected [27]. The recording medium or screen is placed behind the sample. The dark areas on the medium represent defects and the light ones represent no defects or areas of low radiation [27].

4.5.1 Advantages and Disadvantages The method of testing is extremely accurate, sensitive, is a non-contact process and most defects such as porosity can be detected. The method eliminates safety hazards [27]. However, high energy supply for experiment conduction is a challenge [27].

5. Conclusion

The various techniques used in NDT are reviewed. All techniques have equal importance in the field of Inspection and Maintenance. Although some techniques such as Thermography stand out in quality of inspection, others too have their pros and cons and requirements which we have to suitably choose from in order to maximize the quality, quantity and improvement of these techniques. Although some techniques were selectively invented for specific testing of FRP, other techniques were invented to reduce human efforts and provide us with the basics of damage detection. The issue of expenses will always be a major issue in any field and therefore, all these factors especially the application of these techniques should be chosen suitably in order to better the field of inspection and maintenance in FRP.

References

[1] P. Composites, *History of frp*, http://www.mateenbar.com/history-of-frp/(2019)
[2] R.S.Y. Wong, Msc Thesis, University of Toronto (2001)
[3] Cosmos, *How do heat shield on spacecrafts work*, https://cosmosmagazine.com/technology(2016)
[4] Composites-UK, *Bridges*, https://compositesuk.co.uk/composite-materials(2019)
[5] S. X.C, S.R.Hallett, International Journal of Impact Engineering **109**, 178(2017)
[6] U. Polimeno, M. Meo, Composite Structures **91**, 398(2009)
[7] D. Dionysopoulos, G.P. Malfense-Fierro, M. Meo, F. Ciampa, NDT and EInternational **95**, 9 (2018)
[8] P. Böer, L. Holliday, T.H. Kang, Construction and Building Materials **50**, 209(2014)
[9] Worldspec,*Whatisndtanditsbenefits*,https://www.worldspec.org/What-is-NDT-and- its-Benefits/(2019)
[10] BI-NDT, *The relationship between shm, cm and ndt*, http://www.bindt.org/forums/moodys-memoranda/the-relationship-between-shm-cm-and-ndt/ (2013)
[11] TSP, *4 benefits of nondestructive testing*, http://tspndt.com/non-destructive-testing- industrial-supplies-blog/4-benefits-of-nondestructive-testing(2015)
[12] M.P.D.C. YunzeHe, GuiYunTian, Composites: Part B **59**, 196(2014)
[13] H. Kaiser, V. Karbhari, Int. J. Materials and Product Technology, **21(5)**, 349(2004)
[14] M.F. Ali Sophian, Guiyun Tian, Chin. J. Mech. Eng. **30**, 500(2017)
[15] P.H. Eric Kuhn, Emmanuel Valot, Composite Structures **94**, 1155(2012)
[16] A.M. R. de Oliveira, Computers and Structures **86**, 367(2008)
[17] T.M.A.P.D.G.Aggelis,N.-M.Barkoula,CompositesScienceandTechnology**72**,1127 (2012)
[18] K.R.K.A.H.SandeepDegala,PiervincenzoRizzo,ConstructionandBuildingMaterials **23**, 2016 (2009)
[19] Inspectioneering, *Overview of acoustic emission testing (aet)*, https://inspectioneering.com/tag/acousticemission, month
[20] N.R. Center, *Introduction to acoustic emission testing*, https://www.nde-ed.org/EducationResources/CommunityCollege/Other.
[21] T.Asimakopoulou,T.Philippidis,FatigueLifePredictionofCompositesandComposite Structures pp. 466–504(2010)
[22] O.Ley, V.Godinez, *Non-destructive evaluation (NDE) of aerospace composites: ap- plication of infrared (IR) thermography* (Woodhead Publishing Series in Composites Science and Engineering, 2013), pp.309–334
[23] E.G. Henneke, K.L. Reifsnider, W.W. Stinchcomb, Journal of Metals pp. 11–15(1979)
[24] G.M. Carlomagno, C. Meola, NDT & E International **35**, 559(2002)
[25] T. Yu, T.K. Cheng, A. Zhou, D. Lau, Construction and Building Materials **109**, 146 (2016)
[26] C. Goidescu, H. Welemane, C. Garnier, M. Fazzini, R. Brault, E. Péronnet, S. Mistou, Composites: Part B **48**, 95(2013)
[27] ASNT, *Introduction to nondestructive testing*,https://www.asnt.org/minorSiteSections/
[28] N.R. Center, *Basic principles of ultrasonic testing*, https://www.nde-ed.org/EducationResource

IOP Conf. Series: Materials Science and Engineering **520** (2019) 012004 doi:10.1088/1757-899X/520/1/012004

Axisymmetric Multiple Contact Problem for a System of Annular Punches and Foundation with Nonuniform Coating of Variable Thickness

K E Kazakov[1,2,*] **and S P Kurdina**[2]

[1] Ishlinsky Institute for Problems in Mechanics of the Russian Academy of Sciences, Vernadsky Ave. 101 Bldg 1, Moscow, 119526, Russia
[2] Bauman Moscow State Technical University, 2nd Baumanskaya Str. 5/1, Moscow, 105005, Russia

*E-mail: kazakov-ke@yandex.ru

Abstract. Contact problem for a system of rigid annular rough punches and foundation with nonuniform coating of variable thickness is considered. We assume that coating nonuniformity, it's thickness, and forms of the punches can be described by rapidly changing functions. Mathematical model of the axisymmetric problem is compiled. It's efficient solution is constructed by using special Manzhirov projection method. In this solution coating properties and shapes of the contacting surfaces are represented by separate terms and factors. Such a result can hardly be done by other known methods. Qualitative conclusions are presented.

1. Introduction

Contact problems for foundations with coatings have been studied in good number of papers (see, e.g., [1–15]. Various cases were considered: multiple and single contact, bodies with uniform and nonuniform coatings, various profiles of coatings and punches. This paper for the first time provide solution of axisymmetric problem for the common case of arbitrary system of annular rigid punches taking into account coating nonuniformity and shapes of the contacting surfaces.

2. Axisymmetric multiple contact problem

Layer of a thickness h_{lower} made of viscoelastic aging material at moment τ_{lower} coated by elastic layer of variable thickness $h(r)$ lies on a rigid undeformable basis. Modulus of elasticity and Poisson's ratio depends of upper layer depends on the radial coordinate, i.e. $E = E(r)$ and $\nu = \nu(r)$. Such foundations sometimes are called as surface nonuniform foundations. The material nonuniformity usually results from the process of layer by layer manufacturing (additive manufacturing) of a coating (e.g., laser treatment, ion implantation, etc. [16, 17]) and its further processing. We assume that upper layer is soft, i.e. it's modulus of elasticity is less than the modulus of elasticity of the lower layer or they are of the same order of magnitude. There is smooth contact or perfect contact between layers and between the lower layer and the rigid base.

At time $\tau_0 \geq \tau_{\text{lower}}$, the forces $P_i(t)$, applied along the z-axis, starts to indent system of n smooth annular rigid punches, which axes coincides with the z-axis, into the surface of such a foundation. The punch base forms describes by functions $f_i(r)$. The contact regions are constant and bounded by the circles with radii $r = a_i$ and $r = b_i$, where a_i and b_i are inner and outer radii of ith punch, $i = 1,2,\ldots,n$.

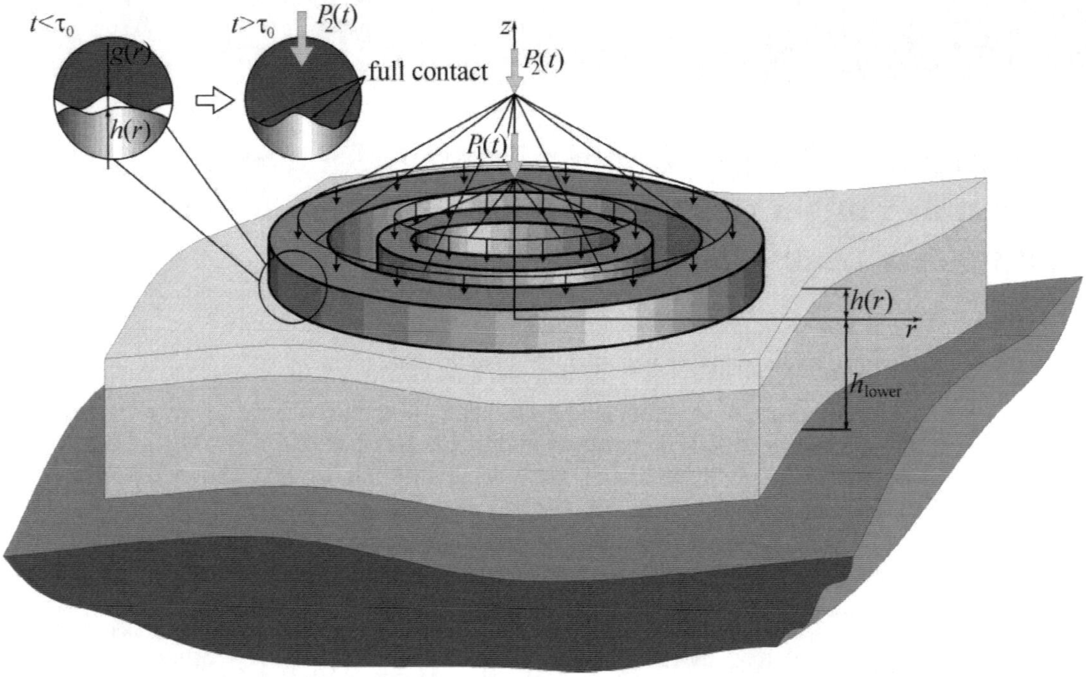

Figure 1. Axisymmetric contact problem for two annular punches.

The coating is assumed to be thin compared with the contact areas, i.e., its thickness satisfies the condition $h(r) \ll \max_{i=1,2,\ldots,n}(b_i - a_i)$. Scheme of contact interaction for 2 punches presented in figure 1.

To derive the mathematical model of the problem, we replace the system of punches by some normally distributed load acting on regions $r \in [a_i,b_i]$, i=1,2,…,n (under the punches), and equal to zero outside these regions. Then the vertical displacement of the upper face of the foundation described above on each region (under each punch) can be written as (see, for example, [18])

$$\frac{q_i(r,t)h(r)}{R(r)} + \frac{2(1-\nu_{\text{lower}}^2)}{h_{\text{lower}}}\sum_{j=1}^{n}\left\{\frac{1}{E_{\text{lower}}(t-\tau_{\text{lower}})}\int_{a_j}^{b_j} k_{\text{ax}}\left(\frac{r}{h_{\text{lower}}},\frac{\rho}{h_{\text{lower}}}\right)q_j(\rho,t)\rho d\rho\right.$$
$$\left.-\int_{\tau_0}^{t}\frac{K(t-\tau_{\text{lower}},\tau-\tau_{\text{lower}})}{E_{\text{lower}}(t-\tau_{\text{lower}})}\int_{a_j}^{b_j} k_{\text{ax}}\left(\frac{r}{h_{\text{lower}}},\frac{\rho}{h_{\text{lower}}}\right)q_j(\rho,\tau)\rho d\rho\right\} = \delta_i(t) - g_i(r), \tag{1}$$

where $q(r,t)$ is unknown contact pressure, $g_i(r) = f_i(r) - h(r) + h_{0i}$ ($h_{0i} = -\min_{r\in[ai,bi]}\}[f_i(r) - h(r)]$) describes distance between contact surfaces in nondeformable state and called backlash function; contact rigidity $R(r)$ depend on the contact conditions between coating and lower layer (see, for example, [19]); in the case of a perfect coating-layer contact, we have $R(r) = E(r)[1 - \nu(r)]/\{[1 + \nu(r)][1 - 2\nu(r)]\}$, and in the case of an smooth contact, $R(r)=E(r)/[1 - \nu^2(r)]$, ν_{lower} and $E_{\text{lower}}(t - \tau_{\text{lower}})$ are the Young modulus and Poisson's ratio of the lower layer; $\delta_i(t)$ is are the punch settlements; $K(t,\tau)$ is tensile creep kernel ($K(t,\tau) = E_{\text{lower}}(\tau)\,\partial[1/E_{\text{lower}}(\tau) + C_{\text{lower}}(t,\tau)]/\partial\tau$, $C_{\text{lower}}(t,\tau)$ is the tensile creep function; $k_{\text{ax}}(r/h_{\text{lower}},\rho/h_{\text{lower}})$ is known kernel of the plane contact problem, which has the form (see, for example, [19, 20]): $k_{\text{ax}}(r,\rho) = \int_0^\infty L(u)J_0(ru)J_0(\rho u)du$, and, in the case of a smooth contact between the lower layer and the rigid base, $L(u) = [\cosh(2u) - 1]/[\sinh(2u) + 2u]$ and in the case of a perfect contact, $L(u) = [2\kappa\sinh(2u) - 4u]/[2\kappa\cosh(2u) + 4u^2 + 1 + \kappa^2]$, $\kappa = 3 - 4\nu_{\text{lower}}$.

We supplement Eq. (1) with the conditions of the punches equilibrium on the foundation

$$\int_{a_i}^{b_i} q_i(\rho,t)\rho\, d\rho = P_i(t), \quad i = 1,2,\ldots,n. \tag{2}$$

There exist several versions of mathematical statements for the axisymmetric contact problem for a system of punches. It is easy to show that there is one of two types of conditions on each punch: the load force is given and the punch settlement is given. Thus, there are three versions of mathematical statements of contact problem for a system of punches in axisymmetric case.

3. Model representation

Let us make the change of variables in (1) and (2) by the formulas

$$(r^*)^2 = \frac{r^2 - a_i^2}{b_i^2 - a_i^2},\quad (\rho^*)^2 = \frac{\rho^2 - a_j^2}{b_j^2 - a_j^2},\quad t^* = \frac{t}{\tau_0},\quad \tau^*_{\text{lower}} = \frac{\tau_{\text{lower}}}{\tau_0},\quad \lambda = \frac{h_{\text{lower}}}{b_1 - a_1},\quad \eta^{i*} = \frac{a_i}{b_1 - a_1},$$

$$(\zeta^i)^2 = \frac{b_i^2 - a_i^2}{(b_1 - a_1)^2},\quad \delta^{i*}(t^*) = \frac{\delta_i(t)\zeta^i}{b_1 - a_1},\quad g^{i*}(r^*) = \frac{g_i(r)\zeta^i}{b_1 - a_1},\quad m^{i*}(r^*) = \frac{E_\circ h(r)}{2(1-\nu^2_{\text{lower}})R(r)(b_1 - a_1)},$$

$$c^*(t^*) = \frac{E_{\text{lower}}(t - \tau_{\text{lower}})}{E_\circ},\quad q^{i*}(r^*, t^*) = \frac{2(1-\nu^2_{\text{lower}})q_i(r,t)\zeta^i}{E_{\text{lower}}(t - \tau_{\text{lower}})},\quad P^{i*}(t^*) = \frac{4P_i(t)(1-\nu^2_{\text{lower}})\zeta^i}{\pi E_{\text{lower}}(t - \tau_{\text{lower}})(b_i^2 - a_i^2)}, \quad (3)$$

$$\mathbf{F}^{ij*} f(r^*) = \int_0^1 k^{ij}(r^*, \rho^*) f(\rho^*)\rho^* \, d\rho^*,\quad k^{ij}(r^*, \rho^*) = \frac{\zeta^i \zeta^j}{\lambda} k_{\text{ax}}\left(\frac{r}{h_{\text{lower}}}, \frac{\rho}{h_{\text{lower}}}\right),$$

$$\mathbf{V}^* f(t^*) = \int_1^{t^*} K^*(t^*, \tau^*) f(\tau^*) d\tau^*,\quad K^*(t^*, \tau^*) = K(t - \tau_{\text{lower}}, \tau - \tau_{\text{lower}})\tau_0,\quad i, j = 1,2,\ldots,n.$$

Having omitted asterisks in the obtained relations, we arrive at the system of integral equations and additional conditions in the form ($r \in [0,1]$, $t \geq 1$)

$$c(t) m^i(r) q^i(r,t) + (\mathbf{I} - \mathbf{V}) \sum_{j=1}^{n} \mathbf{F}^{ij} q^j(r,t) = \delta^i(t) - g^i(r),\quad \int_0^1 q^i(\rho, t)\rho \, d\rho = P^i(t),\quad i = 1,2,\ldots,n. \quad (4)$$

Here I is identity operator. Assuming that

$$\mathbf{q}(r,t) = q^i(r,t)\mathbf{i}^i,\quad \boldsymbol{\delta}(t) = \delta^i(t)\mathbf{i}^i,\quad \mathbf{g}(r) = g^i(r)\mathbf{i}^i,\quad \mathbf{P}(t) = P^i(t)\mathbf{i}^i,$$

$$\mathbf{k}(r,\rho) = k^{ij}(r,\rho)\mathbf{i}^i\mathbf{i}^j,\quad \mathbf{F}\mathbf{f}(r) = \int_0^1 \mathbf{k}(r,\rho)\cdot\mathbf{f}(\rho)\rho d\rho,\quad \mathbf{D}(r) = D^{ij}(r)\mathbf{i}^i\mathbf{i}^j,\quad D^{ij}(r) = \begin{cases} m^i(r) & i = j, \\ 0 & i \neq j, \end{cases} \quad (5)$$

we can represent system with additional conditions (4) as

$$c(t)\mathbf{D}(r)\mathbf{q}(r,t) + (\mathbf{I} - \mathbf{V})\mathbf{F}\mathbf{q}(r,t) = \boldsymbol{\delta}(t) - \mathbf{g}(r),\quad \int_0^1 \mathbf{q}(\rho,t)\rho \, d\rho = \mathbf{P}(t).\quad r \in [0,1],\quad t \geq 1. \quad (6)$$

Hereinafter, it will be the summation over repeated upper indices i and j from 1 to n if the left side of the formula is independent of the index. Note that vector functions D(r) and g(r) connect with coating rigidity, coating width, and punch base forms. These parameters can be described by a rapidly changing function.

In what follows, we construct the solution of the of two-dimensional operator equation with the vector auxiliary conditions (6), which contains integral operators with constant as well as variable limits of integration and $2n$ different rapidly changing functions.

4. Analytical solution

4.1. Special form of the contact pressure

We should see the solution for all versions of the problem in the form

$$\mathbf{q}(r,t) = \tilde{\mathbf{q}}(r,t) - \mathbf{D}^{-1}(r)\cdot\mathbf{g}(r)/c(t). \quad (7)$$

where $\mathbf{D}^{-1}(r) = [D^{ij}(r)]^{-1}\mathbf{i}^i\mathbf{i}^j$ and $\tilde{\mathbf{q}}(r,t)$ is new unknown function. Then operator equation and auxiliary

IOP Conf. Series: Materials Science and Engineering **520** (2019) 012004 doi:10.1088/1757-899X/520/1/012004

conditions (6) can be reduced to the following equations:

$$c(t)\mathbf{D}(r)\tilde{\mathbf{q}}(r,t)+(\mathbf{I}-\mathbf{V})\mathbf{F}\tilde{\mathbf{q}}(r,t)=\boldsymbol{\delta}(t)-\tilde{c}(t)\tilde{\mathbf{g}}(r),\quad \int_0^1\tilde{\mathbf{q}}(\rho,t)\rho\,d\rho=\tilde{\mathbf{P}}(t),\quad r\in[0,1],\quad t\geq 1,\qquad (8)$$

where

$$\tilde{\mathbf{g}}(r)=\int_0^1\mathbf{k}(r,\rho)\cdot\mathbf{D}^{-1}(\rho)\cdot\mathbf{g}(\rho)\rho d\rho,\ \tilde{c}(t)=-(\mathbf{I}-\mathbf{V})c^{-1}(t),\ \tilde{\mathbf{P}}(t)=\mathbf{P}(t)+c^{-1}(t)\int_0^1\mathbf{D}^{-1}(\rho)\cdot\mathbf{g}(\rho)\rho d\rho.\quad (9)$$

We obtain new mixed operator equation with n rapidly changing function $m_i(r)$ ($i = 1,2,\ldots,n$) supplemented by two vector conditions. Last term in right-hand side of operator equation (8) is "good": its smoothness defined by kernel $\mathbf{k}(r,\rho)$.

We will construct the solution of equations (7) for the version when all forces are known.

4.2. Special basis and solution representation

By introducing in (8) notations $\tilde{\mathbf{q}}(r,t)=\mathbf{D}^{-1/2}(r)\cdot\mathbf{Q}(r,t)$, $\mathbf{k}(r,\rho)=\mathbf{D}^{1/2}(r)\cdot\mathbf{K}(r,\rho)\cdot\mathbf{D}^{1/2}(\rho)$, $\mathbf{Gf}(r)=\int_0^1\mathbf{K}(r,\rho)\cdot\mathbf{f}(\rho)\rho d\rho$ we obtain new operator equation and additional conditions

$$c(t)\mathbf{Q}(r,t)+(\mathbf{I}-\mathbf{V})\mathbf{G}\mathbf{Q}(r,t)=\mathbf{D}^{-1/2}(r)\cdot[\boldsymbol{\delta}(t)-\tilde{c}(t)\tilde{\mathbf{g}}(r)],\quad \int_0^1\mathbf{D}^{-1/2}(r)\mathbf{Q}(\rho,t)\rho\,d\rho=\tilde{\mathbf{P}}(t).\quad (10)$$

Here $\mathbf{D}^{1/2}(r)=\sqrt{D^{ij}(r)\mathbf{i}^i\mathbf{i}^j}$, $\mathbf{D}^{-1/2}(r)=\sqrt{1/D^{ij}(r)\mathbf{i}^i\mathbf{i}^j}$; $\mathbf{Q}(r,t)$ is new unknown function, and $\mathbf{K}(r,\rho)$ is new kernel.

We seek the solution of Eq. (10) in the class of vector functions continuous in time t in the Hilbert space $L_2([0,1],V)$ (see [21]). To this end, we at first construct an orthonormal system of vector functions in $L_2([0,1],V)$ which contains the factors $1/\sqrt{m^i(r)}$ and remaining basis functions can be written as the products of vector functions depending on r and weight functions $1/\sqrt{m^i(r)}$. To this end we will orthonormal on [0,1] the linearly independent system of vector-functions $\{\mathbf{i}^i r^k/\sqrt{m^i(r)}\}_{i=1,2,\ldots,n,\,k=0,1,2,\ldots}$ by the formulas:

$$\mathbf{p}_k^i(r)=\mathbf{D}^{-1/2}(r)\cdot\mathbf{p}_k^{i\circ}(r),\quad \mathbf{p}_k^{i\circ}(r)=p_k^{i\circ}(r)\mathbf{i}^i,\quad d_{-1,i}=1,\quad J_{k,i}=\int_0^1\frac{\rho^{2k+1}d\rho}{m^i(\rho)},$$

$$d_{k,i}=\begin{vmatrix}J_{0,i}&\cdots&J_{k,i}\\ \vdots&\ddots&\vdots\\ J_{k,i}&\cdots&J_{2k,i}\end{vmatrix},\quad p_k^{i\circ}(r)=\frac{1}{\sqrt{d_{k-1,i}d_{k,i}}}\begin{vmatrix}J_{0,i}&J_{1,i}&\cdots&J_{k,i}\\ \vdots&\vdots&\ddots&\vdots\\ J_{k-1,i}&J_{k,i}&\ddots&J_{2k-1,i}\\ 1&r^2&\cdots&r^{2k}\end{vmatrix},\quad \begin{matrix}i=1,2,\ldots,n,\\ k=0,1,2,\ldots,\\ r\in[0,1].\end{matrix}\qquad (11)$$

Thus, $\mathbf{p}_k^i(r)$ ($i = 1,2,\ldots,n$, $k = 0,1,2,\ldots$) is a basis in $L_2([0,1],V)$.

4.3. Solution of the problem

If Hilbert space $L_2([0,1],V)$ represents as the direct sum of the Euclidean space $L_2^{(0)}([0,1],V)$ with the basis $\{\mathbf{p}_0^i(r)\}_{i=1,2,\ldots,n}$ and Hilbert space $L_2^{(1)}([0,1],V)$ with the basis $\{\mathbf{p}_1^i(r),\mathbf{p}_2^i(r),\mathbf{p}_3^i(r),\ldots\}_{i=1,2,\ldots,n}$ then the integrand and the right-hand side of integral equation (10) can be presented as the algebraic sum of functions continuous in time t and ranging in $L_2^{(0)}([0,1],V)$ and $L_2^{(1)}([0,1],V)$, respectively, i.e.,

$$\begin{gathered}\mathbf{Q}(r,t)=\mathbf{Q}_0(r,t)+\mathbf{Q}_1(r,t),\quad \mathbf{D}^{-1}(r)\cdot[\boldsymbol{\delta}(t)-\tilde{c}(t)\tilde{\mathbf{g}}(r)]=\boldsymbol{\Delta}_0(r,t)+\boldsymbol{\Delta}_1(r,t),\\ \mathbf{Q}_0(r,t)=z_0^i(t)\mathbf{p}_0^i(r),\quad \boldsymbol{\Delta}_0(r,t)=\mathbf{D}^{-1/2}(r)\cdot[\boldsymbol{\delta}(t)-\tilde{c}(t)\tilde{\mathbf{g}}_0(r)]=\Delta_0^i(t)\mathbf{p}_0^i(r),\\ \Delta_0^i(t)=\sqrt{J_{0,i}}\,\delta^i(t)-\tilde{g}_0^i\tilde{c}(t),\quad \Delta_m^i(t)=-\tilde{g}_m^i\tilde{c}(t),\quad i=1,2,\ldots,n,\quad m=1,2,3,\ldots,\end{gathered}\qquad (12)$$

here $\tilde{\mathbf{g}}(r)=\tilde{\mathbf{g}}_0(r)+\tilde{\mathbf{g}}_1(r)$, $\mathbf{Q}_0(r,t)$, $\Delta_0(r,t)$, and $\tilde{\mathbf{g}}_0(r)$ belongs to $L_2^{(0)}([0,1],V)$ and $\mathbf{Q}_1(r,t)$, $\Delta_1(r,t)$, $\tilde{\mathbf{g}}_1(r)$ belongs to $L_2^{(1)}([0,1],V)$. It is easy to show that coefficients $\tilde{g}^i_m$ has a form

$$\tilde{g}^i_m=\sum_{l=0}^{\infty}K^{ij}_{ml}\int_0^1\frac{p_l^{j\circ}(\rho)g^j(\rho)}{m^i(\rho)}\rho d\rho,\quad i=1,2,\dots,n,\quad m=0,1,2,\dots \tag{13}$$

where K^{ij}_{ml} are expansion coefficients of the kernel $\mathbf{K}(r,\rho)$:

$$K^{ij}_{ml}=\int_0^1\frac{k^{ij}(r,\rho)p_m^{i\circ}(r)p_l^{j\circ}(\rho)}{m^i(r)m^j(\rho)}r\rho drd\rho,\quad i,j=1,2,\dots,n,\quad m,l=0,1,2,\dots \tag{14}$$

The formula for $\mathbf{Q}(r,t)$ contains known term $\mathbf{Q}_0(r,t)$ which is determined by the auxiliary conditions (10): $z_0^i(t)=\tilde{P}^i(t)/\sqrt{J_{0,i}}$. Conversely, for the right-hand side, one should find $\Delta_0(r,t)$, while term $\Delta_1(r,t)$ is known and determined be the function $\tilde{\mathbf{g}}_1(r)$. These peculiarities permit one to class the resulting problem as a specific case of the generalized projection problem stated in [21–23].

Having introduced the orthoprojectors $\mathbf{P}_0$ and $\mathbf{P}_1$ mapping the space $L_2([0,1],V)$ onto subspaces $L_2^{(0)}([0,1],V)$ and $L_2^{(1)}([0,1],V)$, respectively, we will apply the orthoprojector $\mathbf{P}_1$ to operator equation (10). As a result, we obtain the equation for determining $\mathbf{Q}_1(r,t)$ with a known right-hand side

$$c(t)\mathbf{Q}_1(r,t)+(\mathbf{I}-\mathbf{V})\mathbf{P}_1\mathbf{G}\mathbf{Q}_1(r,t)=\mathbf{\Delta}_1(r,t)-(\mathbf{I}-\mathbf{V})\mathbf{P}_1\mathbf{G}\mathbf{Q}_0(r,t). \tag{15}$$

We will construct its solution in the form of an expansion in the eigenfunctions of the operator $\mathbf{P}_1\mathbf{G}$. The spectral problem for this operator can be written in the form

$$\mathbf{P}_1\mathbf{G}\boldsymbol{\varphi}_k(r)=\gamma_k\boldsymbol{\varphi}_k(r),\quad \boldsymbol{\varphi}_k(r)=\sum_{m=1}^{\infty}\psi^i_{km}\mathbf{p}^i_m(r),\quad k=1,2,3,\dots \tag{16}$$

This problem leads to the search for solutions of spectral problem about coefficients $_k$ and ψ^i_{km} (i = 1,2,…,n, k,m = 1,2,3,…): $\sum_{l=1}^{\infty}K^{ij}_{ml}\psi^j_{kl}=\gamma_k\psi^i_{km}$. We expand the all functions in equation (15) with respect to the new basis functions $\varphi_k(r)$ (k = 1,2,3,…) in $L_2^{(1)}([0,1],V)$, i.e.,

$$\mathbf{Q}_1(r,t)=\sum_{k=1}^{\infty}z_k(t)\boldsymbol{\varphi}_k(r),\quad \mathbf{P}_1\mathbf{G}\mathbf{Q}_0(r,t)=\sum_{k=1}^{\infty}\sigma_k(t)\boldsymbol{\varphi}_k(r),\quad \mathbf{\Delta}_1(r,t)=\sum_{k=1}^{\infty}\Delta_k(t)\boldsymbol{\varphi}_k(r), \tag{17}$$

where functions $_k(t)$ and $_k(t)$ has a form $\sigma_k(t)=\sum_{m=1}^{\infty}K^{ij}_{m0}\psi^i_{km}z_0^j(t)$, $\Delta_k(t)=\sum_{m=1}^{\infty}\psi^i_{km}\Delta^i_m(t)$.

Substituting these representations into (11) we obtain formula for the unknown functions $z_k(t)$

$$z_k(t)=(\mathbf{I}+\mathbf{W}_k)\frac{-(\mathbf{I}-\mathbf{V})\sigma_k(t)+\Delta_k(t)}{c(t)+\gamma_k},\quad \mathbf{W}_kf(t)=\int_1^t R_k^*(t,\tau)f(\tau)d\tau,\quad k=1,2,3,\dots, \tag{18}$$

where $R_k^*(t,\tau)$ is the resolvent of the kernel $K_k^*(t,\tau)=\gamma_kK(t,\tau)/[c(t)+\gamma_k]$.

Note that the solution has a form (i = 1,2,…,n)

$$q^i(r,t)=\frac{1}{m^i(r)}\left[z_0^i(t)p_0^{i\circ}(r)+\sum_{k=1}^{\infty}z_k(t)\sum_{m=1}^{\infty}\psi^i_{km}p_m^{i\circ}(r)\right]-\frac{g^i(r)}{c(t)m^i(r)},\quad r\in[0,1],\quad t\geq 1, \tag{19}$$

i.e., one can explicitly write out the functions $m_i(r)$ and $g_i(r)$ in the solution. The coating nonuniformity and shapes of the contacting surfaces which can described by complicated and rapidly changing functions are related to these functions in the relations of change of variables. The formulas obtained permit obtaining efficient analytical solutions.

5. Conclusions

Axisymmetric multiple contact problem for foundation with nonuniform coating and arbitrary system of annular punches is posed. The corresponding mathematical model is given and analyzed. Possible variants of the problem statement are formulated. The analytical solution for the version of known forces is obtained using generalized projection method. In relation of contact stresses (19) coating nonuniformity and shapes of the contacting surfaces are represented by separate terms and factors. It is allows one to perform effective computations for actual nonuniformities and punch shapes and analyze carefully the behavior of the punches on the layer, taking into account the complex properties and forms of the bodies and the mutual influence of the punches.

Acknowledgments
This work was financially supported in part by the Russian Foundation for Basic Research under grant No. 18-01-00770-a.

References

[1] Manzhirov A V 1983 *J. Appl. Math. Mech.* **47**(4) 558–66 doi: 10.1016/0021-8928(83)90098-9
[2] Manzhirov A V 1985 *J. Appl. Math. Mech.* **49**(6) 777–82 doi: 10.1016/0021-8928(85)90016-4
[3] Aleksandrov V M and Manzhirov A V 1987 *J. Appl. Mech. Techn. Phys.* **28**(5), 781–6 doi: 10.1007/BF00912035
[4] Chen Y F, Erdogan F 1996 *J. Mech. Phys. Solids* **44**(5) 771–87 doi: 10.1016/0022-5096(96)00002-6
[5] Pauk V and Zastrau B W 2004 *J. Theor. Appl. Mech.* **42**(1) 107–24
[6] Kul'chyts'kyi-Zhyhailo R and Rogowski G 2009 *Mater. Sci.* **45**(6) 845–58 doi: 10.1007/s11003-010-9251-y
[7] Kazakov K E 2009 *J. Phys.: Conf. Ser.* **181** 012013 doi: 10.1088/1742-6596/181/1/012013
[8] Vasiliev A, Volkov S, Aizikovich S, and Jeng Y-R 2014 *ZAMM* **94**(9) 705–12 doi: 10.1002/zamm.201300067
[9] Aizikovich S M, Vasil'ev A S, and Volkov S S 2015 *J. Appl. Math. Mech.* **79**(5) 500–5 doi: 10.1016/j.jappmathmech.2016.03.011
[10] Ledesma-Alonso R, Raphaël E, Léger L, et al. 2016 *Proc. Math. Phys. Engng Sci.* **472**(2193) 20160235 doi: 10.1098/rspa.2016.0235
[11] Kazakov K E, Kurdina S P, and Manzhirov A V 2017 *Proc. IUTAM* **23** 201–9 doi: 10.1016/j.piutam.2017.06.021
[12] Kazakov K E and Manzhirov A V 2017 *Mech. Solids* **52**(4) 444–51 doi: 10.3103/S0025654417040112
[13] Manzhirov A V and Kazakov K E 2018 *Math. Models Comput. Simul.* **10**(3) 314–21 doi: 10.1134/S2070048218030109
[14] Kazakov K E 2018 *AIP Conf. Proc.* **2053** 040041 doi: 10.1063/1.5084479
[15] Kazakov K E and Kurdina S P 2018 *J. Phys.: Conf. Ser.* **991** 012040 doi: 10.1088/1742-6596/991/1/012040
[16] Gibson I, Rosen D W, and Stucker B 2010 *Additive Manufacturing Technologies: Rapid Prototyping to Direct Digital Manufacturing* (New York, Springer Verlag, LLC)
[17] Gebhardt A and Hoetter J S 2016 *Additive Manufacturing: 3D Printing for Prototyping and Manufacturing* (Munich, Carl Hanser Verlag)
[18] Arutyunyan N Kh and Manzhirov A V 1999 *Contact Problems of Theory of Creep* (Erevan, Izdat. Inst. Mekaniki NAN RA) [in Russian]
[19] Alexandrov V M and Kovalenko E V 1986 *Problems in Mechanics of Continuous Media with Mixed Boundary Conditions* (Moscow, Nauka) [in Russian]
[20] Vorovich I I, Alexandrov V M, and Babeshko V A 1974 *Nonclassical Mixed Problems in the Theory of Elasticity* (Moscow, Nauka) [in Russian]
[21] Manzhirov A V 2016 *Dokl. Phys.* **61**(10) 489–93 doi: 10.1134/S1028335816100025
[22] Manzhirov A V 2016 *Lect. Notes Engng Comp. Sci.* **2224**(1) 10–5
[23] Manzhirov A V 2017 Mixed integral equations and their applications to mechanics *IAENG Transactions on Engineering Sciences* (Netherlands, World Scientific Pub Singapore) pp 8–21 doi: 10.1142/9789813226203_0002

IOP Conf. Series: Materials Science and Engineering **520** (2019) 012005 doi:10.1088/1757-899X/520/1/012005

Sealing System Analysis of Rotary Piston Pump

M Li[1, 3], X Zhang[1], W Yang[1, a], J Zhang[2, 3] and Z Zhou[1]

[1] Shandong University, Research Center of Geotechnical and Structural Engineering, 250061 Jinan, China
[2] Beihang University, School of Transportation Science and Engineering, Beijing Key Laboratory for Cooperative Vehicle Infrastructure Systems and Safety Control, 100191 Beijing, China
[3] Chalmers University of Technology, Department of Architecture and Civil Engineering, SE412 96 Gothenburg, Sweden

[a] Corresponding author : weimin-yang@163.com

Abstract. Rotary piston pump is designed by the Wankel engine. The structure, working feature and sealing system of rotary piston pump are introduced. The mathematical model of the internal cylinder profile and the theoretical flow are obtained by analysing the structural characteristics of the new pump. The axial seal and face seal of the rotary piston pump and their structures are analysed. The force analysis of the sealing strip is carried out in various cases. It can be concluded from the analysis that, for axial sealing, the key parameters that affect seal strip wear are working pressure, rotate speed of the crankshaft and the friction coefficient between sealing strip and inner wall of the cylinder. For axial sealing, the force of the spring under the curved sealing strips, the size of the sealing rings and curved sealing strips and material of them are the key design parameters.

1. Introduction
In 1957, NSU's engineer Paschke proposed the original design of Wankel engine. In 1963, Wankel introduced the principle of the engine in detail [1].Although many manufacturers have tried to adopt the Wankel engine, only Mazda began to use this new type of engine in its automobile products since 1962 [2].Similar to engine, Wankel's principle can be applied to the other systems. For example, the Rankine Cycle Expander, Badr[3] presents a steam plant power cycle system using Wankel expander. Like Wankel engine and Wankel expander, the rotary piston pump (RPP) offers a greater output density than piston pump, has higher specific power [4], and avoids the imbalance of inertia moment and inertial force caused by the reciprocating working mode of the piston, which causes the inertial load of the support member in the piston pump to be intensified and easily causes unstable vibration and noise of the whole device [5, 6]. Because of the good performance, RPP has been applied in some fields, such as artificial heart [7] and grouting [8]. However, the poor sealing system is still the key point which restricts the spread of Wankel engine and RPP.

2. Rotary piston pump

2.1. Structure of RPP

The main parts of single-cylinder RPP include a crankshaft, a rotor, a cylinder block, a gear seat, a bearing block and two cover plates, as shown in Figure 1. The gear seat is in the centre of the cover plate. The crankshaft includes a main journal and a connecting journal. The centre of the cross section of the main journal is concentric with the centre of the cylinder block, and the centre of the cross section of the connecting journal is concentric with the centre of the rotor. The gear form gear seat meshes with the annular gear in the rotor, and the gear ratio is 2:3. The internal cylinder profile is a double arc outer trochoidal, and the rotor divides the cylinder into three working chambers. The eccentric rotary motion of the rotor makes the volume of the three working chambers change constantly, so as to complete the suction and discharge.

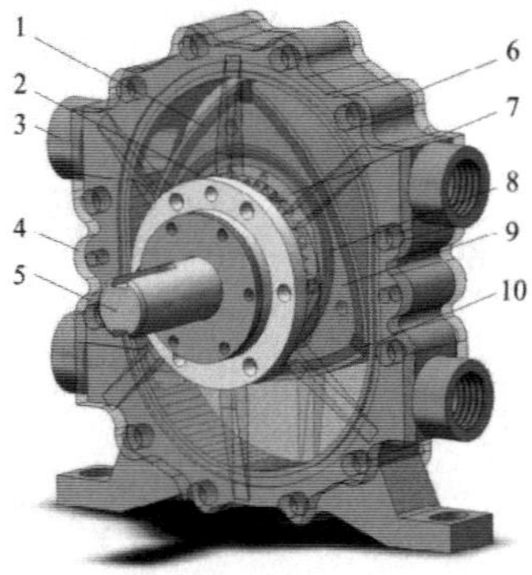

1- Curved sealing strip, 2-gear seat, 3-inlet or outlet, 4-cover plate, 5-crankshaft, 6-cylinder block, 7-annular gear, 8-sealing ring, 9-rotor, 10-sealing strip

Figure 1. Structure of RPP

2.2. Working feature

The internal cylinder profile is also the envelope of the rotor vertices' moving trajectory[9,10]. In the X-Y coordinate system of Figure 2, the distance from the rotor center to the vertex is O_2B, which value is R, and the distance between the rotor center and the crankshaft's main journal center is O_1O_2, which value is e. When the rotor rotaries from Figure 2(a) to Figure 2(b), the motion angle of the point B is β, and the motion angle ogre point M is φ. Based on the regulation of the rotor motion, there are two equal curve appear: $\alpha \cdot r_2=\beta \cdot r_1$. From the gear radio of the two meshed gears, $\alpha/\beta=r_2/r_1=2/3$ can be worked out, then $\beta=\alpha+\varphi=3\varphi$. From the geometric relationship in Figure 3, the internal cylinder profile can be obtained:

$$\left.\begin{aligned} X &= e\cos\beta + R\cos\frac{\beta}{3} \\ Y &= e\sin\beta + R\sin\frac{\beta}{3} \end{aligned}\right\} \quad (1)$$

The annular gear is fixed, and drives the internal cylinder profile to move by the gear on gear seat, forming three envelopes, which are the sides of rotor. There are three equivalent arcs to be used as the sides of the rotor. AC whose dot is G is one of them, and its radius GA is

$$\mathrm{GA} = \sqrt{\mathrm{GH}^2 + \mathrm{HA}^2} = \frac{(R-e)^2 + 3e^2}{R-4e} \quad (2)$$

The areas S_{ADF} can be expressed by Equation (1):

$$S_{\mathrm{ADF}} = \int_{\pi}^{\frac{2\pi}{3}} (e\sin 3\varphi + R\sin\varphi)(e\cos 3\varphi + R\cos\varphi)' d\varphi \quad (3)$$

The areas S_{AEF} can be expressed by the conditions known:

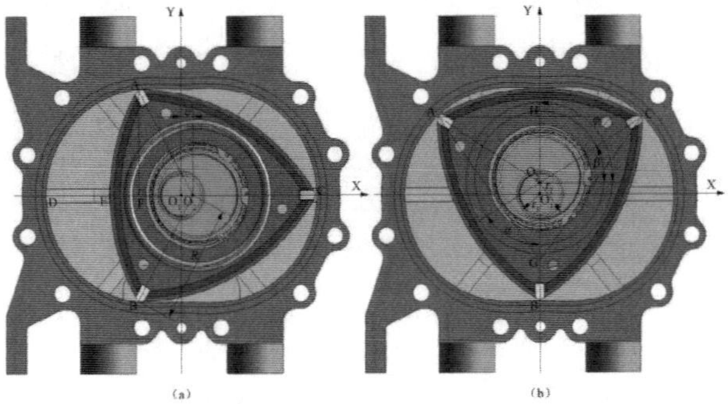

Figure 2. The geometric model of the RPP

$$S_{\text{AEF}} = (\frac{\pi}{4} - \frac{3\sqrt{3}}{8})R^2 \tag{4}$$

In Figure 2(a), the working chamber of ADBE gets the largest volume, and the area S_{ADBE} is

$$S_{\text{ADBE}} = 2S_{\text{ADE}} = 2(S_{\text{ADF}} - S_{\text{AEF}}) = \pi e^2 + \frac{3\sqrt{3}eR}{2} - \left(\frac{\pi}{6} - \frac{\sqrt{3}}{2}\right)R^2 \tag{5}$$

Finally, the theoretical flow rate (q_0) of the RPP is

$$q_0 = 2nsS_{\text{ADBE}} = \left(4s\pi e^2 + 6\sqrt{3}eRs - \frac{2\pi}{3}R^2 s + 2\sqrt{3}R^2 s\right)\text{n} \tag{6}$$

The three working chambers (W-C) alternate in their suction and discharge. The volume of a W-C changes with the rotary angle φ of the crankshaft, as shown in Figure 3. S. Li [11] studied the fluctuation process of flow rate and pressure in the W-Cs and found that flow and pressure fluctuated regularly, and the fluctuation amplitude was large.

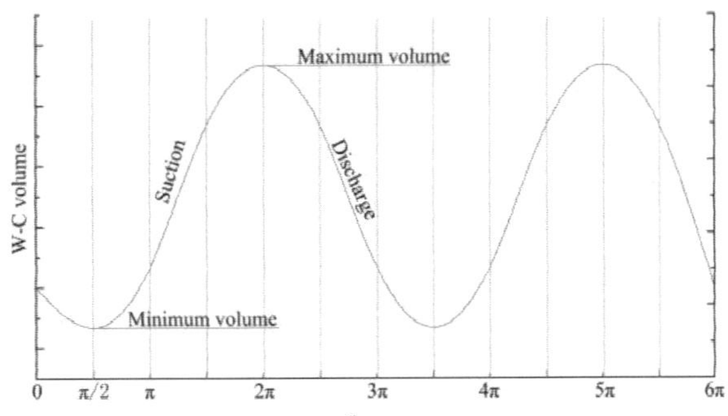

Figure 3. The volume of a W-C changes with the rotary angle φ of the crankshaft

3. Sealing system

3.1. Sealing structure

Compared with the Wankel engine, the liquid pressure in RPP is higher than the gas pressure in engine. In addition to this, the liquid which RPP pumps can corrode the rotor faces and the internal faces of the cover plates. These lead to the worse working conditions. Therefore, the structure, size and material of the sealing system are important. If the parameters are not designed properly, there will be deformation and severe wear, as shown in Figure 4.

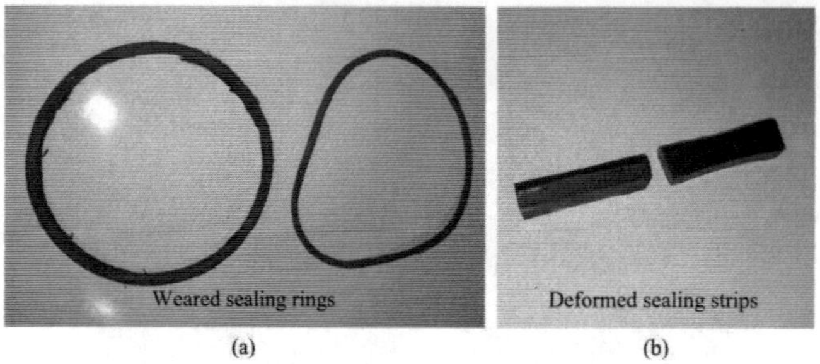

Figure 4. Wear and deformation of sealing rings and strips

3.2. Axial sealing

The axial sealing strips between working chambers are designed to prevent leakage between the each chamber. With the fluctuation of the flow and pressure, the sealing strips swing back and forth in the grooves. When one of the sealing strips comes to the position A in Figure 5(a), the differential pressure and rotor rotation have the same direction. The force analysis is shown in Figure 5(b).

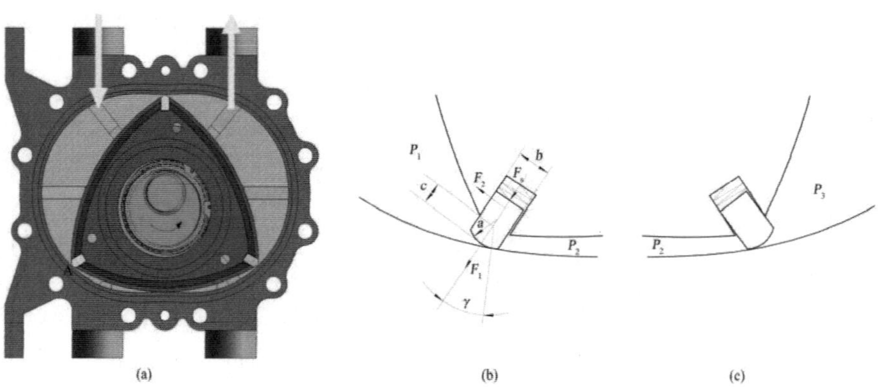

Figure 5. Analysis of the sealing strip

(1) Friction can be ignore

If the material of the sealing strip is hard enough to ignore the friction, like the Wankel engine, F_2 can be expressed as [12]:

$$F_2 = lhP_2 - lcP_1 + m\omega^2 e\sin\frac{2}{3}\varphi \tag{7}$$

And F_1 can be expressed as:

$$F_1 = lbP_2 - P_2\left(\frac{b}{2} - a\sin\gamma\right) - P_1\left(\frac{b}{2} + a\sin\gamma\right) + m\omega^2\left(\frac{r}{9} + e\cos\frac{2}{3}\varphi\right) + F_s \tag{8}$$

Where, l is the length of the sealing strip. h is the height of the sealing strip. P_1 and P_2 are pressures of working chambers. c is the length of the sealing strip outside the groove. m is the mass of the sealing strip. ω is the angular velocity of the crankshaft. b is the width of the sealing strip. a is the radius of the sealing strip head. γ is the contact angle. r is the length of the sealing strip from the center of rotor. F_s is the force of spring.

The hard materials that are available are alloy cast iron, iron titanium sintered alloy and hot press silicon nitride. When the sealing strip comes to the other side, and the differential pressure and rotor rotation have the opposite direction, the sealing strip will incline in the groove because the opposite forces, as shown in Figure 5(c). F_1 and F_1 become:

$$\left.\begin{aligned} F_1 &= lbP_2 - P_2\left(\frac{b}{2} - a\sin\gamma\right) - P_3\left(\frac{b}{2} + a\sin\gamma\right) + m\omega^2\left(\frac{r}{9} + e\cos\frac{2}{3}\varphi\right) + F_s \\ F_2 &= lhP_2 - lcP_3 + m\omega^2 e\sin\frac{2}{3}\varphi \end{aligned}\right\} \tag{9}$$

Where, P_3 is the pressure of another working chamber.

(2) Friction cannot be ignore

When the pressure required is less than 0.7Mpa, rigid sealing strip is feasible. But if the pressure is greater than 0.7Mpa, flexible sealing material is required, and the friction of the sealing strip cannot be ignored. The hard materials that available are polytetrafluoroethylene containing carbon fiber, high strength nylon and rubber. Under the influence of pressure F_N, the sealing strip is subject to the friction F_μ in the opposite direction of motion, as shown in Figure 6.

F_N can be expressed as:

$$F_N = F_1\cos\gamma - F_2\sin\gamma \tag{10}$$

Then, the friction F_μ can be obtained:

$$F_\mu = \mu\left(\frac{b\cos\gamma l}{2} + al\sin\gamma\cos\gamma + lh\sin\gamma - lcP_1\sin\gamma\right)(P_2 - P_1) + m\mu\omega^2\left(\frac{r\cos\gamma}{9} + e\cos\gamma\cos\frac{2}{3}\varphi - e\sin\gamma\sin\frac{2}{3}\varphi\right) + \mu F_s\cos\gamma \tag{11}$$

In conclusion, the key parameters that affect seal strip wear are working pressure, rotate speed of the crankshaft and the friction coefficient between sealing strip and inner wall of the cylinder.

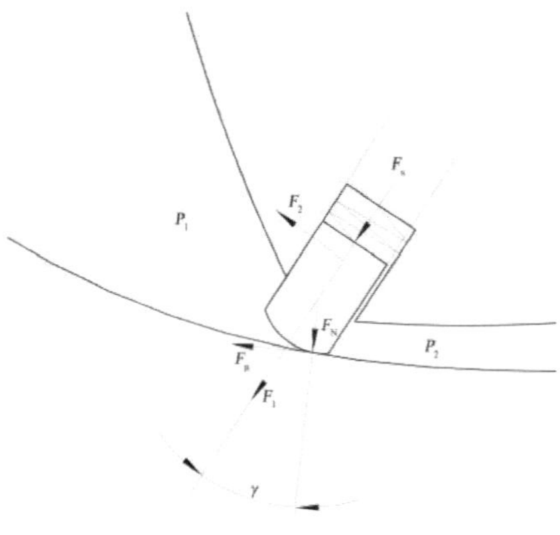

Figure 6. Friction analysis of the sealing strip

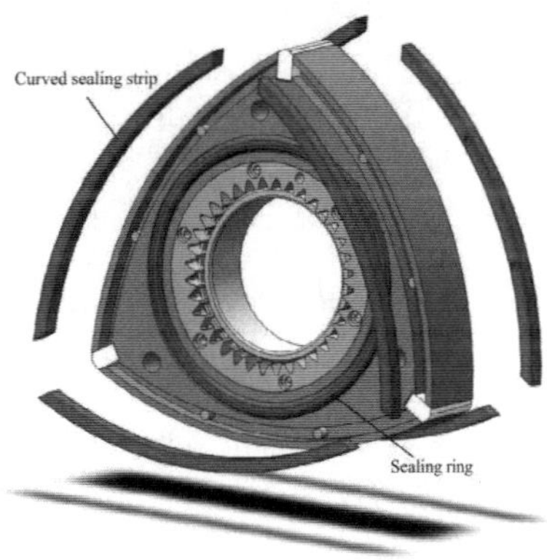

Figure 7. Radial sealing

3.3. Radial sealing

The function of the radial sealing, also called end seal, is to prevent the liquid from the working chamber to leak to the end face of the rotor, shown in Figure 7. The radial sealing includes curved sealing strips, sealing rings, and the axial end face of sealing strips. The curved sealing strips work by the spring at the bottom of grooves and the pressure of the liquid in working chambers. The sealing rings work by the self-elasticity and the extrusion of cover plates. The curved sealing strips and sealing rings are two seals to make sure there's no leak. The force of the spring under the curved strips, the size of the sealing rings and curved sealing strips and material of them are the key design parameters. However, as the friction force and inertia force of all points on the curved sealing strips and sealing rings are different, it's difficult to generalize the relationship between each parameter and the degree of wear. Therefore, each parameter can only be quantified through a large number of experiments.

IOP Conf. Series: Materials Science and Engineering **520** (2019) 012005 doi:10.1088/1757-899X/520/1/012005

4. Conclusion

Based on the Wankel engine, a rotary piston pump is designed. The working principle, structure and double arc outer trochoidal model of RPP are introduced in detail. This article focuses on the components of the RPP axial seal and face seal and their structures. The force analysis of the sealing strip is carried out in various cases. It can be concluded from the analysis that, for axial sealing, the key parameters that affect seal strip wear are working pressure, rotate speed of the crankshaft and the friction coefficient between sealing strip and inner wall of the cylinder. For axial sealing, the force of the spring under the curved sealing strips, the size of the sealing rings and curved sealing strips and material of them are the key design parameters.

References

[1] Wankel F 1963 *Rotary piston machines* (London: Liffe Books)

[2] Ozcanli M, Bas O, Akar M A, Yildizhan S and Serin H 2018 Experimental and theoretical analysis of Wankel engine performance *International Journal of Hydrogen Energy* March 04

[3] Badr O and Naik S, O'Callaghan P W and Probert S D 1991 Wankel engines as steam expanders: design considerations *Appl. Energ.* **40** 157

[4] Ansdale R F 1969 *The Wankel RC engine* (Cranbury: Barnes and Co)

[5] Pennock G R and Beard J E Force analysis of the apex seals in the wankel rotary compressor including the influence of fluctuations in the crankshaft speed *Mechanism and Machine Theory* **32** 349

[6] Lee C H, Jiang K C, Jin P and Prewett P D 2004 Design and fabrication of a micro Wankel engine using MEMS technology *Microelectron. Eng.* **73-74** 529

[7] Wappenschmidt J, Autschbach R, Steinseifer U, Schmitz-Rode T, Margreiter R, Klima G and Goetzenich 2016 A Rotary piston blood pumps: past developments and future potential of a unique pump type *Expert Rev. Med. Devic.* **13** 759

[8] Li S, Li M, Zhang X and Zhang Q 2018 Study on A New Design of Grouting Pump for Managing Water Inrush in Karst Tunnels *The Civil Engineering Journal* **27** 357

[9] Li M, Li S, Zhang X, Zhang Q, Wang Z and Hao P 2019 Performance evaluation and flow analysis of two-cylinder triangular rotor pump based on experiment and numerical simulation *Adv J. Mech. Des. Syst* **13** 0003

[10] Li M, Li S, Zhang X and Zhang Q 2019 Experimental investigations on extrusive eccentric rotary pump *Journal of Harbin Institute of Technology* **51** 154

[11] Li S, Li M, Zhang X, Zhang Q, Hao P and Wang Z 2018 Investigation on Performance of Triangular Rotor Pump Based on Numerical Simulation and Experiment *Transactions of the Chinese Society for Agricultural Machinery* **49** 389

[12] Lu F and Yu N 1990 *Triangle rotary piston engine*(Beijing: National Defense Industry Press)

Economic Evaluation of Seawater Cooling System Applying Flooded Type Evaporator

Jung-In Yoon[1, a], Chang-Hyo Son[1], Choon-Geun Moon[1], Joon-Hyuk Lee[1], Kwang-Seok Lee[1], Suk-Ho Jung[1] and Doo-Yeong Kwak[1]

[1] Pukyong National University, South Korea

[a] yoonji@pknu.ac.kr; Tel: +82-51-629-6180

Abstract. The ministry of oceans and fisheries of South Korea promoted the modernization of the purse seine vessels, and the seawater cooling system was proposed as part of the modernization. In the previous study, heat transfer characteristics were analysed with respect to material of tubes to offer fundamental design data of seawater cooling system applying flooded type evaporator by Yoon et al. In this paper, the economic evaluation of the seawater cooling system for comparison with the cooling method using crushed freshwater ice was performed with the net present value method. The initial investment cost of the seawater cooling system was calculated by 181 thousand dollars as the total cost of equipment and facility. The annual benefit was calculated by 47 thousand dollars which subtracts the maintenance cost of the seawater cooling system from the cost of ice and salt. Consequently, the net present value of the seawater cooling system presented positive when the discount rate is lower than 21.4%, which means the seawater cooling system can be considered economical. Moreover, the discount rate has 4.5% as the social discount rate in South Korea, so the net present value of the seawater cooling system has 154 thousand dollars approximately.

1. Introduction

The purse seine fishery supplies 90% of catch of mackerel in South Korea. The purse seine fishery is a method of catching fishes by lighting up at night and spreading the net, and the purse seine fishery consists of six vessels, which are one main vessel, two lighting vessels and three fish carriers. The ministry of oceans and fisheries in South Korea promoted the modernization of the purse seine vessels. Because of the modernization, the developed fishing vessels will be reduced from six to four vessels, which are one main vessel, one lighting vessel and two fish carriers, and it improved crew's welfare of the fishing vessel as well. Moreover, seawater cooling system is proposed as a part of the modernization. Cooling method using freshwater ice damages on the catch due to the sharp crystal of the ice and change of salinity in the fish tank. But the seawater cooling system make seawater supplied into the fish tank be cooled directly, so it can solve the decline of the catch's quality.

Gorgy. E et al. [1] focused on the pool boiling performance of both R123 and R134a on smooth and enhanced tubes. As results of this research, for smooth tube testing the heat flux ranger were 7.3~130.7kW/m^2 and 7.5 ~ 60.7kW/m^2 for R134a and R123, respectively; with resulting heat transfer coefficient ranges of 1,798.9 ~ 11,379W/m^{2o}C and 535.4~3,181.8W/m^{2o}C.

Kang et al. [2] analysed heat transfer characteristic of aluminium-brass tube using R-134a. As a result of this research, increase rate of heat transfer factor was decreased rapidly when flow velocity of

IOP Conf. Series: Materials Science and Engineering **520** (2019) 012006 doi:10.1088/1757-899X/520/1/012006

seawater was faster than 1.8m/s. Furthermore, overall heat transfer coefficient was decreased 60% approximately at low temperature comparing with high temperature.

Yoon et al. [3] analysed heat transfer characteristics of flooded type evaporator using R22 and R134a. The variables were flow velocity of seawater is changed from 0.6m/s ~ 2.1m/s, and inlet temperature of seawater is changed from 5°C to 15°C, respectively. As results of the experiment, the heat transfer coefficient of R22 had higher than R134a and the low-fin tube had a higher heat transfer coefficient than the plane tube in above conditions.

Yoon et al. [4] analysed heat transfer characteristics with respect to material of tubes applying flooded type evaporator in seawater cooling system. The material used aluminium-brass and copper-nickel to enhance heat transfer performance, and the results of the analysis presented that the aluminium-brass low-fin tube had higher heat transfer coefficient than copper-nickel low-fin tube.

Economic evaluation is as important as evaluation of the technological validity of the seawater cooling system. Economic evaluation is as important as evaluation of the technological validity of the seawater cooling system.

Hong et al. [5] used net present value, internal rate of return and cost-benefit ratio method to evaluate economical validity of the development of automatic operation system for hair tail trolling line in Jeju region, South Korea. Also, the net present value was calculated using social discount rate, 4.5%.

Bae et al. [6] used net present value method, internal rate of return and cost-benefit ratio to evaluate economic feasibility and impact analysis on energy transformation project of macro-algae biomass. Also, this was considered economical when the discount rate is lower than 15.7%.

In this paper, the economic evaluation of the seawater cooling system for comparison with the cooling method using crushed freshwater ice was performed with the net present value method. When economic evaluation was conducted, experimental results in the previous study of the seawater cooling system was used [4].

2. Seawater cooling system

2.1. Heat transfer characteristics of flooded type evaporator

Prior to designing the seawater cooling system for fishing vessel with 163kW or 238kW cooling capacity, seawater cooling system with 15kW was developed to provide system performance analysis and fundamental design data. The effect of material and the refrigerant R-22, R-134a on the heat transfer promoting tube in the flooded type evaporator used for the seawater cooling system was analysed to acquire fundamental data of design and heat transfer enhancement. When analysing of heat transfer characteristics, inlet temperature of the cooling seawater was set to 10°C, evaporating temperature was set to 0°C, and flow velocity was set to 1.8m/s.

2.1.1. Experimental apparatus Figure 1 presents experimental apparatus of the flooded type evaporator and schematic diagram. The apparatus is composed of the flooded type evaporator, open type reciprocating compressor, condenser, receiver, expansion valve, oil recovery heat exchanger, surge drum, separator and oil return collector. Working principle is as follows. Gaseous refrigerant from the compressor outlet flows into the condenser, and the working fluid is condensed by exchanging the heat with cooling water. Liquid refrigerant at the outlet of the condenser is expanded through expansion valve, and evaporates at the evaporator by acquiring heat from the seawater. Evaporated refrigerant flows into the compressor, and it circulates.

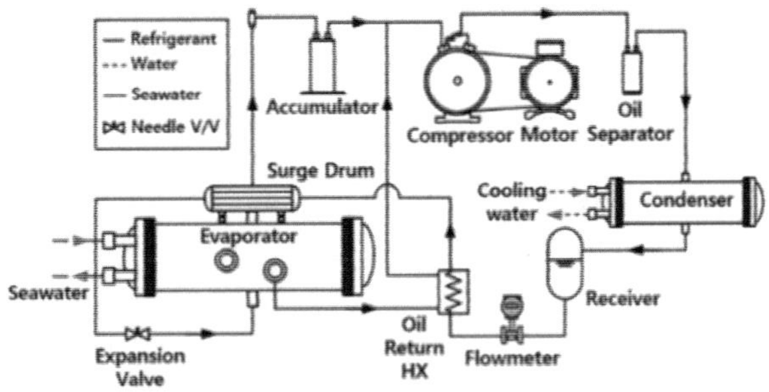

Figure 1. The schematic diagram of the apparatus to analysis heat transfer characteristics [4]

2.1.2. Experimental results As shown in figure 2, heat transfer coefficient tended to increase when evaporating temperature decreased. In addition, the heat transfer coefficient increased 76.6% as the evaporating temperature decreased from 5°C to -5°C. It was considered that increment of the evaporating capacity was larger than that of the mean temperature difference. Besides, heat transfer coefficient of R22 is 0.35kW/m^{2}K higher than R134a with same material of the tube. While, aluminium-brass had 0.25kW/m^{2}K higher coefficient than the copper-nickel tube with same refrigerant used.

The above discussion about the experimental result on heat transfer characteristics was referred in prior research [4].

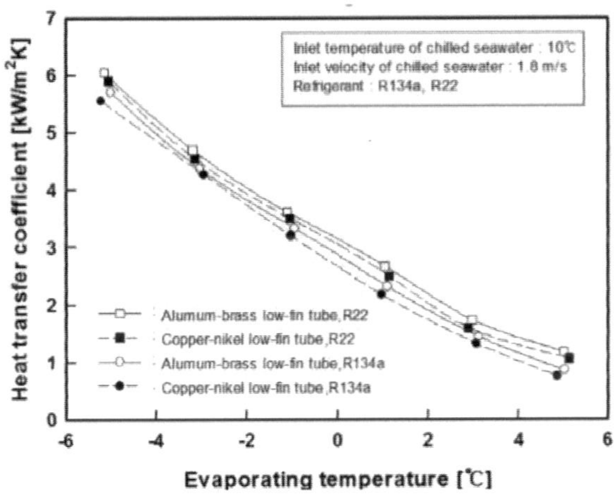

Figure 2. The experimental results of heat transfer coefficient based on evaporating temperature [4]

2.2. Performance analysis of flooded type evaporator

2.2.1 Experimental apparatus In order to analyse the efficiency of seawater cooling system, experimental apparatus was installed on land. Figure 3 shows schematic diagram and experimental apparatus of the seawater cooling system. It was composed of the steam boiler, fish tank, control panel, cooling tower, heat exchanger, invertor, data acquisition equipment and the seawater cooling system. The inverter controlled the flow rate of the seawater, and cooling tower and steam boiler controlled temperature of that for making similar condition with vessel. Working principle is as follows. In the seawater cooling system, gaseous refrigerant from the compressor outlet into the condenser, and the working fluid is condensed by exchanging the heat with cooling water. Liquid refrigerant is collected in receiver and it is exchanged heat with refrigerant from the evaporator outlet in super heater. The refrigerant at the outlet of the super heater is expanded by expansion valve, and evaporates at the evaporator by acquiring heat from seawater. Evaporated refrigerant flows into the super heater, and it circulates. In blue line, divaricated water from the evaporator outlet into fish hold heater, and the water is heated by exchanging with boiling water to maintain temperature steady. In addition, the water is mixed with else water from evaporator, and then flows into the fish hold tank. In the sky-blue line, water is cooled in cooling tower and divaricated water is heated in seawater heater to maintain temperature steady. Also, in the red line water which flows into the seawater heater or fish hold heater was heated with boiler.

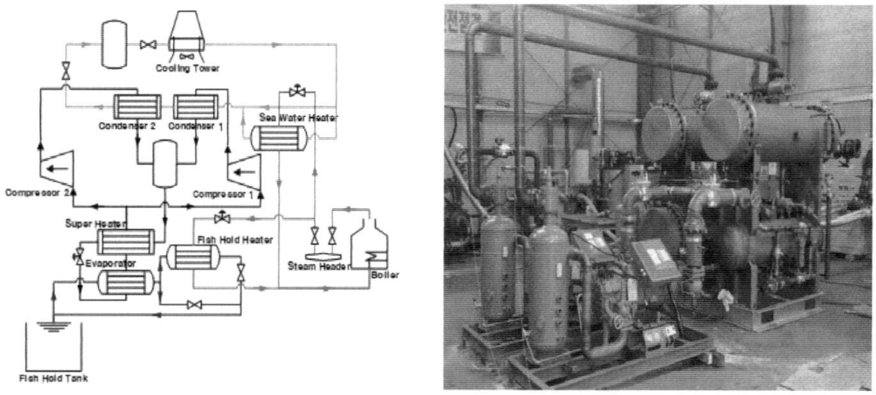

Figure 3. The schematic diagram and experimental apparatus of the seawater cooling system

2.2.2. Experimental results Table 1 shows the performance results of the 238kW seawater cooling system. The performance experiment of the system was conducted with the condition of flow rate of the seawater, 90m3/h. Steady state was considered when the seawater temperature maintains 3°C, which is the storage temperature for the mackerel.

Table 1. Experimental results of the performance analysis

Parameter	Unit	Value
Evaporator duty	kW	240.3
Pump power consumption	kW	50
Compressor power consumption	kW	72
Total power consumption	kW	122
Coefficient of performance	-	3.34

3. Economic evaluation

As a method of economic evaluation, the net present value method was applied. Durable years of the system was set to 10 years, and the discount rate had 4.5% as the social discount rate. The discount rate means rate to evaluate present value having same value with future value.

3.1. Net present value method

Net present value means sum between the net benefit calculated as the present value for set period and the present value. For instance, if the net present value has positive value, it can be considered economical. The net present value is calculated using equation (1).

$$NPV = -I + \sum_{r=1}^{n} \frac{C}{(1+i)^r} \tag{1}$$

In the equation (1), NPV means net present value, I is the initial investment cost, C is the benefit, i is the discount rate and r is durable years.

3.2. Conditions of economic evaluation

Target vessel of economic evaluation has used 70,000 kg of ice, 367 kg of salt per a fishing, and annual cost of the ice and the salt was calculated by 4 cents per kg, and 54.5 cents per kg investigated in South Korea. In addition, machine price and facilities are estimated by 130 thousand dollars and 51 thousand dollars respectively. The annual maintenance cost was calculated by 2% of the initial investment cost, 3,620 dollars, and operation cost is calculated by 12,357 dollars. These costs were calculated as follows.

Heat of fusion of the ice was divided into the evaporating capacity of the system, 201kW, then operation time was calculated by 81 hours. Vessels operate mainly in the 50-70% load range of the electric generator. The total power consumption was 125kW referred to Table 1, so the required power was calculated 137.5kW including the safety factor of 10%. The fuel consumption was investigated to be 36.2 litre per hour at 50% output of a diesel generator model having a rated output of 275kW [7]. Therefore, it was calculated as 31,280 dollars considering annual fuel consumption of 61,576.2L per year, considering the average value of 50.8 cents per litre of monthly diesel for 2017 provided by Korea National Oil Corporation. Table 2 shows the purchase cost of crushed freshwater ice and salt and the initial investment cost and maintenance and operation cost of the seawater cooling system. In addition, benefit is 47,023 dollars which subtracts the maintenance cost of the seawater cooling system from the cost of ice and salt.

Table 2. Conditions of economic evaluation

Parameter	Cost (dollars)
Initial investment cost	181,000
Annual cost of crushed freshwater ice and salt	63,000
Annual maintenance and operation cost	15,977
Benefit	47,023

3.3. Results of economic evaluation

The net present value was calculated according to the discount rate, and the result is shown in figure 4. Consequently, the net present value of the seawater cooling system presents positive value when the discount rate is lower than 21.4%. It means the seawater cooling system is considered economical. Moreover, the discount rate has 4.5% as the social discount rate in South Korea, so the net present value of the seawater cooling system has 154 thousand dollars approximately.

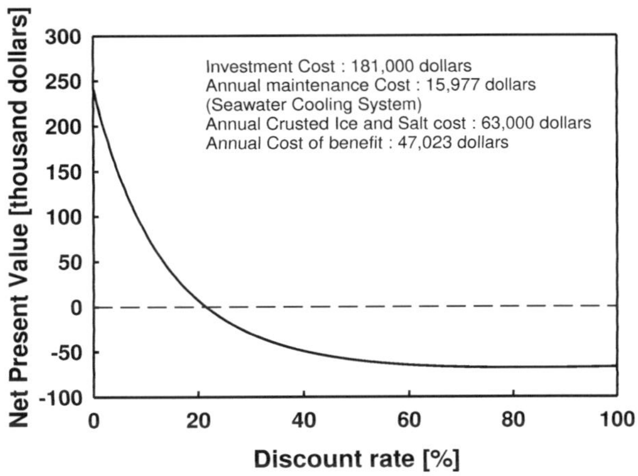

Figure 4. Net present value of the discount rate

4. Conclusions

Prior to evaluate economic feasibility of seawater cooling system, heat transfer characteristics analysis and efficiency analysis gone along. In this paper, seawater cooling system was analysed heat transfer coefficients and evaluated economics based on that experimental results using the net present value. The summary of the results is as follows.

Heat transfer coefficient of R22 is 0.35kW/m^{2}K higher than R134a with same material of the tube, and aluminium-brass had 0.25kW/m^{2}K higher coefficient than the copper-nickel tube with same refrigerant used.

The net present value presented negative when the discount rate is larger than 21.4%. The net present value was 154 thousand dollars approximately when the discount rate has 4.5% as the social discount rate in South Korea.

Acknowledgements

This research was a part of the project titled 'Introduction study of seawater cooling system for fishing-ships', funded by the Ministry of Oceans and Fisheries, Korea (20150480)

References

[1] Gorgy E and Eckels S 2012 Local Heat Transfer Coefficient for Pool Boiling of R-134a and R-123 on Smooth and Enhanced Tubes *International Journal of Heat and Mass Transfer* **55** pp. 3021-3028

[2] Kang I H, Seol S H, Yoon J I and Son C H. 2017 Evaporating heat transfer characteristics of Aluminum-brass tube for seawater cooling system using R-134a *Journal of the Korean Society of Marine Engineering* **41** pp 197-201

[3] Yoon J I, Son C H, Choi K H, Jung S H, Han H M and Seol S H 2018 Heat Transfer Characteristics of Flood-type Evaporator Using R22 and R 134a for Fishing Vessels *Heat Transfer Engineering* (Online)

[4] Yoon J I, Son C H, Han H M, Lee K S, Lee J M, I and You I D 2018 Heat Transfer Characteristics of Flooded Type Evaporator for Seawater Cooling System *MATEC Web of Conferences* **167** 02014

[5] Hong S W, Yang U G, Kim M K, Park Y S, Park K I and Kim D H 2018 An economic feasibility analysis of the automatic operation system development for hairtail trolling line in Jeju region, Korea *Journal of the Korean Society of Fisheries and Ocean Technology* **54** pp 164-172

[6] Bae J H, Jung H Y and Kim M J 2015 Economic Feasibility and Impact Analysis on Energy Transformation Project of Macro-Algae Biomass *The Korean Society for New And Renwable Energy* **11** pp 29-38

[7] B. H. Lee 2016 A Study on Methodology of Optimal Operation of BESS and Diesel Generators in a Microgrid Considering Efficiency Characteristics According to the Power Ratios of diesel Generators *The Transactions of the Korean Institute of Electrical Engineers* **65** pp 539-546

IOP Conf. Series: Materials Science and Engineering **520** (2019) 012007 doi:10.1088/1757-899X/520/1/012007

Development of a Microfluidic Device to Encapsulate Isocyanate for Autoreactive and Ecological Adhesives

Mariana Costa[1], José P Dias[1], Isabel Pinho[2], Mónica V Loureiro[3], Ana C Marques[3] and Ricardo Simoes[1,a]

[1] Instituto Politécnico do Cávado e do Ave – IPCA, Campus do IPCA, Barcelos, Portugal
[2] CIPADE- Indústria e Investigação de Produtos Adesivos, SA. Av. 1° de Maio, 518 Z. Industrial 1, Apt. 82, 3701-909 São João da Madeira, Portugal
[3] CERENA, Departamento de Engenharia Química, Instituto Superior Técnico, Universidade de Lisboa, Avenida Rovisco Pais, 1049-001 Lisboa, Portugal

[a] Corresponding author: rsimoes@ipca.pt

Abstract. Present study was performed to explore a microfluidic approach for production of microcapsules (MCs) with polyurethane/polyurea shells containing isophorone diisocyanate (IPDI) in the core, for adhesive formulations. The MCs were produced, in a continuous mode, by a system involving a microfluidic device, containing a T-junction where is originated a monodisperse oil-in-water emulsion, followed by interfacial polymerization at the emulsion droplets surface. The resulting MCs are intended to be added to an adhesive base, leading to an autoreactive and ecological monocomponent adhesive, with high potential application. The MCs' morphology and size distribution were evaluated by scanning electron microscopy (SEM) and the results obtained by Fourier transformed infrared spectroscopy (FTIR) revealed a successful encapsulation of IPDI. The advances achieved in the current study will be a potential contribution to the innovation and development of new production methods and eco-friendly products.

1. Introduction

One of the current challenges in science is the development of safer new materials/processes for both human health and the environment, in order to promote sustainability and the improvement of the living conditions of future generations [1,2].

Microencapsulation is a process by which individual particles or droplets of an active agent can be stored within a shell, surrounded or coated with a continuous film of polymeric material. This is a growing field that has found application in many areas such as food, medicine, pesticides, environmental and biological engineering, cosmetics and coatings [3]. Moreover, microcapsules (MCs) were also used in adhesive, which could control release of additives and adhesives at the right time and the right place and improve the compatibility of certain ingredients to broaden their application fields [4].

Encapsulation of specific chemicals can be used to protect active agents from detrimental conditions (heat, light, humidity and exposure to other substances over their lifetime), shield an irritating smell, act as a carrier system, control the rate and moment at which the encapsulated compound is released, prevent the evaporation of volatile compounds and reduce the toxicity of certain active substances [5]. Nowadays, the most reactive encapsulations are performed in batch processes to produce MCs. However, in this kind of processes, because of the highly variable shear used, the sizes and morphologies of the resultant microcapsules can vary enormously [6].

Microfluidics offers an alternate and versatile route to overcoming the limitations of batch processes. In microfluidic devices, the drops are produced in series, contrary to bulk emulsification methods where drops are formed in parallel under the influence of an external shear, which led to a greater control over size and polydispersity of the emulsified drops compared to bulk emulsification methods [7]. In addition, this technology offers other advantages such as reducing the consumption of reagents/samples to pico-liters; decreasing the reaction time to seconds; reducing waste generation; allowing rapid diffusion, mass and heat transfer [7, 8, 9].

Since isocyanates have a good performance in terms of bonding property, water resistance, aging-resistant performance and fine manufacturability, it has been widely used in the field of coatings and adhesive [10]. There are, however, several safety issues as well as potential risks to the environment and human health associated with the use of isocyanate adhesives, in particular when it comes to the direct contact of workers with isocyanate. Therefore, there is an urge to develop a green approach that still allows the use of isocyanate, but without exposure to it.

The goal of this study was the development of a microfluidic device to synthesise MCs with polyurethane/polyurea shell containing isophorone diisocyanate (IPDI) in the core as a self-curing agent. The IPDI was encapsulated through interfacial polymerization between a mixture of IPDI and a methylene diphenyl diisocyanate (MDI) pre-polymer, via stabilized oil-in-water (O/W) emulsion with gum Arabic as a prediluted emulsifier in the aqueous phase, in a microfluidic device. The resulting MCs are intended to be added to an adhesive base.

The proposed method has advantages of being readily controlled, cost-effective and easy to operate, together with its ability to produce a narrow MCs size distribution. By microencapsulating the reactive agent, the product is safer for handling by the industry operators, and the activation mechanism can be controlled more precisely

2. Materials and methods

2.1. Reagents

The chemicals used in this study include methylene diphenyl diisocyanate, MDI, isophorone diisocyanate, IPDI, gum Arabic, lime and glycerol. MDI prepolymer was obtained from CIPADE, IPDI was obtained from Acros Organics, lime was obtained by Lacrilar, gum Arabic and glycerol were obtained from Fisher Chemical. All chemicals were used in the experiments without further purification unless otherwise specified.

Arabic Gum was intended not only to stabilize the emulsion by its action as surfactant, but also to react with the isocyanate as a precursor to the wall of the MCs. The aqueous phase was prepared in a 250 mL beaker, at room temperature, and the solution was stirred in a magnetic stirrer plate during 1h. MDI prepolymer was vigorously mixed with IPDI to yield the oil phase. In some experiments (2nd formulation) glycerol and lime were used to decrease the quantity of gas produced in the reaction, being added to the aqueous solution.

2.2. Microfluidic Device

In the microencapsulation process there are 5 mains steps (Figure 1): the first one consists in the preparation of the phases (aqueous and organic) and then a peristaltic pump will pull the reagents that are inside of microtubes, up to the T-junction, where the emulsion is formed. After this, MCs will be formed during their passage in a plastic coil that is inside a thermal bath. Then, the last step, filtration, occurs.

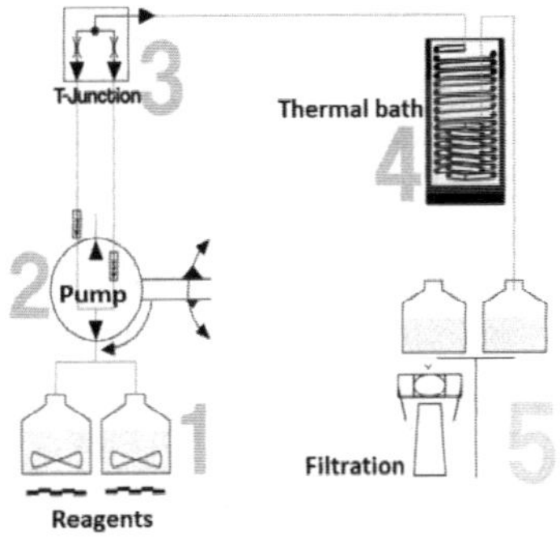

Figure 1. Steps of the microencapsulation system.

The microencapsulation system developed in this work is composed of the following components: (1) heating plates that maintain the reagents homogeneous and at the required temperature; (2) a peristaltic pump which has advantages such as no contamination, low maintenance needs, ability to handle a large variety of fluids, versatility and ease of cleaning ; (3) a T-junction (Figure 2), produced by 3D printing, using a EDEN 260V 3D printer with 600 dpi precision and 0.016 mm layer thickness, with a photocurable polymer as the building material and a water-soluble support material; (4) a thermal bath, set at 60ºC, including an immersed serpentine tube, wherein the MCs flow; (5) a filtration system, composed by a kitasato, a funnel and a vacuum pump.

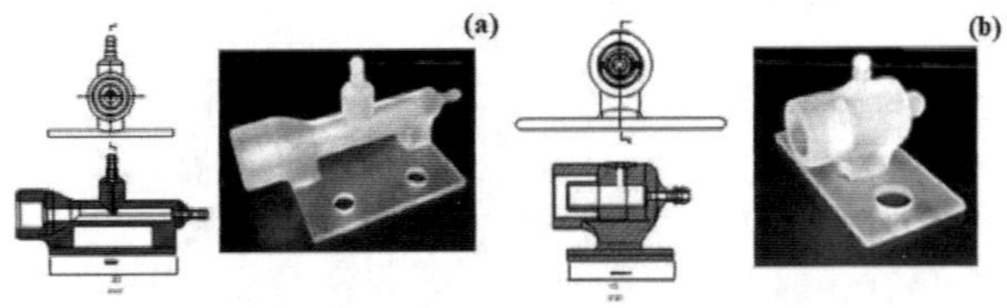

Figure 2. Initial (a) and final (b) concept of T-junction.

In Figure 2 it's possible to see the evolution of T-junction structure. The improvements in relation to the initial concept go through the last one occupied less space, presents better connections with the respective tubes and allows the production of smaller MC's. In T-junction the aqueous phase comes horizontally while the organic phase comes from above, and this is the point where the emulsion's formation starts.

2.3. Characterization

The characterization of the obtained MCs was performed using different techniques such as optical microscopy, scanning electron microscopy (SEM) and Fourier transformed infrared spectroscopy (FTIR).

Morphology observation. SEM analysis was carried out to evaluate the morphology and roughness of the obtained MCs. SEM images were collected in a JEOL JSM7001F (JEOL, Tokyo, Japan) microscope with a FEG-SEM (Field Emission Gun) system. Before the analysis, the MCs were placed in a sample holder using conductive adhesive tape with double face. Afterwards, a conductive Au/Pd thin film was used to coat the samples, through sputtering, using a Quorum Technologies sputter coater model Q150T ES. The MCs obtained were also used to evaluate the MCs' size distribution using the software ImageJ.

FTIR spectroscopy analysis. FTIR analyses were used in order to verify the functional groups present in the obtained MCs. The spectrum was analyzed, and the main peaks were identified (absorbance as a function of wave number). This analysis was performed in a PerkinElmer, Spectrum Two, FTIR spectrometer, equipped with a Pike Technologies MIRacle® 189 Attenuated Total Reflectance (ATR) accessory. The spectra were obtained with a resolution of 4 cm^{-1} and data collection of 8 scans, over a frequency range of 400-4000 cm^{-1}.

3. Results and Discussion

O/W isocyanate containing MCs were produced in a microfluidic device using IPDI/MDI prepolymer in the O phase and water/gum Arabic in the W phase.

In the first experiments performed (1st formulation), the presence of gas bubbles both in the serpentine and inside the MCs was noticed, leading to the MCs agglomeration inside the serpentine and even to the MCs bursting due to the expansion of the gas present in its interior. After a careful observation of the MCs inside the serpentine, it is believed that small bubbles of gas were already formed inside the MCs at the T-Junction, probably due to the residual air dissolved in the W phase or the reaction of the isocyanate with the W phase, forming urea and the release of CO_2 (g). This phenomenon was not observed in the batch process, since the shear stress due to the high-speed agitation releases CO_2 (g) into the atmosphere. In the microfluidic system, the reaction occurs in a confined way, within the serpentine, preventing the CO_2 escape. In order to minimize the formation of gas bubbles during the MCs synthesis, the W phase composition was optimized, by adding lime and glycerol (2nd formulation). The use of lime (CaO) has two purposes: (1) remove CO_2 bubbles (g) and (2) produce fine calcite that may diffuse as a charge on the MCs PU walls and increase its impermeability, through "filtering control" mechanisms [11].

3.1 Effect of the addition of glycerol and lime in the aqueous (W) phase

Figure 3 shows a comparison of the obtained MCs before and after the problem of gas bubbles was solved. As it is possible to confirm, without the gas formation it was possible to obtain smaller and spherical MCs, with a narrow size distribution.

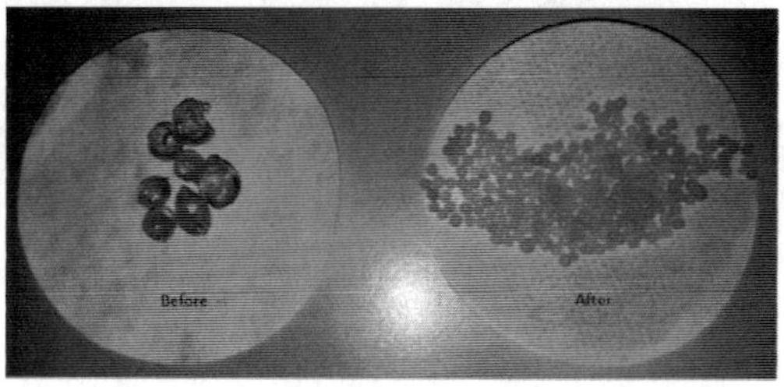

Figure 3. MCs obtained in 1st (left) and 2nd formulation (right).

3.2 Characterization of the obtained MCs

A morphological analysis of the MCs was accessed through SEM, whose photomicrographs are shown in Figure 4. The MCs are found to exhibit a spherical shape, however with some irregularities. Some of the MCs were tried to be crushed, in order to access the MCs morphology, however it was not possible, probably due to the high stiffness and/or thickness of the MCs wall (shell).

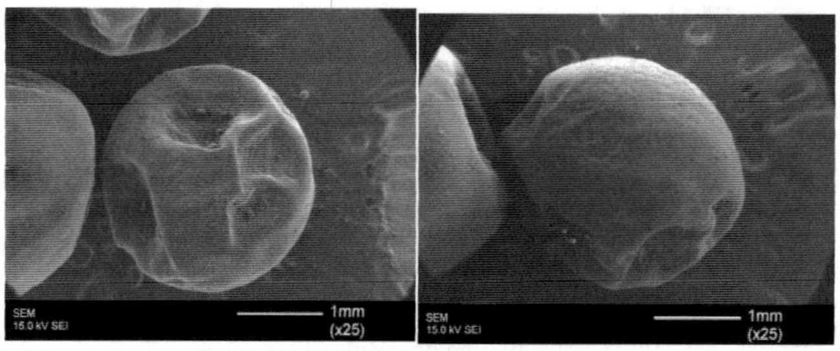

Figure 4. SEM images of MCs.

Through ImageJ software, it was made an analysis of the size distribution of the MCs obtained by 2nd formulation, concluding that the particles have an average diameter of 1100±90 μm. The size of the

MCs is herein found to be bigger than desirable, so it is necessary to adjust the O and W flows in order to obtain smaller MCs. Quevedo et al. [12] studied the effect of aqueous flow rate on capsule size by holding the organic disperse flow rate constant and by varying the aqueous flow rate. They concluded that capsule size gradually decreased with increased aqueous flow rate and, hence, with increasing Reynolds number. So, the next step is performing the experiments with a higher W flow rate.

Samples of MCs obtained by microfluidic device were characterized by FTIR spectroscopy analysis and compared with the spectrum obtained from similar MCs synthesized by conventional emulsion/interfacial polymerization (batch process), under the same conditions, in another lab, by other members of the team. Figure 5 shows the respective spectra. In general, all the spectral features characteristic of polyurethane, polyurea and isocyanate materials are present in the three analysed samples.

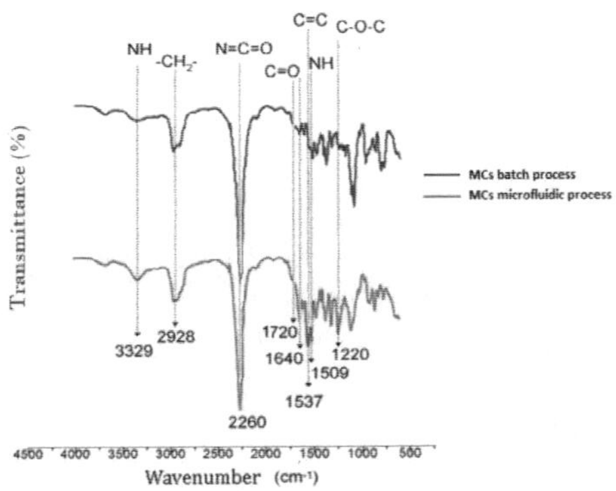

Figure 5. Comparison of FTIR spectra of MCs obtained in batch and microfluidic processes, with identification of the main peaks.

From the FTIR spectra, shown in Figure 5, it is possible to observe an intense band peaked at ca. 2260 cm^{-1}, related to the N=C=O bond stretching vibration, which indicates the presence of unreacted NCO groups in the MCs, confirming a successful encapsulation of IPDI. The peaks ascribed to the presence of amine groups can be observed at 3329 cm^{-1} from N-H stretching of the amine bonds and at 1509 cm^{-1} from N-H bending. Concerning the bands at 2928 and 1537 cm^{-1}, the first one is attributed to the asymmetric stretching vibration of the $-CH_2$ group (aliphatic bonds of MDI or IPDI) while the second is attributed to the symmetrical elongation vibration of the -CH (phenyl groups of MDI). The bands at 1640 and 1210 cm^{-1} correspond to a C=O group from urea and a C-O-C group from urethane. The band at ca. 1720 cm^{-1} corresponds to the C=O group from urethane moieties. The exhibited FTIR spectra reveal the formation of a polyurethane/polyurea MC´s shell and an effective isocyanate (IPDI) encapsulation.

A relative measure of the isocyanate's encapsulation efficiency was calculated, considering the peak at ca. 2260 cm-1 assigned to NCO stretching (wavenumbers in the range of 1926 to 2444 cm^{-1}) and the area of the peak at 1300 cm^{-1} assigned to C-O stretching, typically related to the PUa / PU shell material, which does not tend to suffer significant changes over time. The equation 1 was applied to obtain the relative encapsulation yield, considering the referred areas, previously calculated through the OriginPro 9 software.

$$Y = \frac{Area_{\ 2260\,cm^{-1}}}{Area_{\ 1300\,cm^{-1}}} \tag{1}$$

Where Y is a relative encapsulation yield, which represents an indirect measure of the encapsulation efficiency. The Area 2260 cm^{-1} and Area 1300 cm^{-1} correspond to the areas of the peak related to the NCO group and to the C-O stretch, respectively. The Y values of the MCs obtained by the batch process and by microfluidic process are 47,6 and 19,6, respectively. The MCs synthetized by batch process have shown more encapsulated IPDI.

4. Conclusions and ongoing work

This research provides a new method for synthesizing monodisperse isocyanate MCs. The microfluidic device allows the fabrication of the microcapsules one at a time and offer robust and precise control over the external dimensions.

Since the obtained MCs are bigger than desirable, it is necessary to perform the experiences with a high W flow rate, in order to obtain MCs with a small diameter. Through the characterization by FTIR it was possible to conclude the efficiency of IPDI's encapsulation.

The generated MCs have a potential use in different areas, namely aeronautic and automobile industries, where they are expected to release the isocyanate under specific and controlled conditions, such as pressure or/and temperature. Thus, the use of a microfluidic device for producing isocyanate MCs, may be considered as an important path for both academic and industrial applications.

The incorporation of obtained MCs in adhesives will be studied. Further studies are still in need, regarding the MCs properties and the test of different configurations for the T-junction, in order to reach optimized MCs.

Acknowledgments

This research was developed in the framework of the Co-Promotion Project # 17930, "ECOBOND—Development of new ecological, self-reactive, monocomponent adhesives", which is funded by the Regional Operational Program of Lisbon—LISBOA2020, in the scope of Portugal2020, through the European Regional Development Fund—FEDER.

References

[1] Serrano E, G. Rus and García-Martínez J.2009 *Renew Sust Energ Rev*, **13**, 2373-84
[2] Tundo P, Anastas P, Black D S, Breen J, Collins T, Memoli S, Miyamoto J, Polyakoff M and Tumas W 2000 *Pure Appl. Chem* **72** (7) 1207-28
[3] F. Salaün 2016 *Active Coatings for Smart Textiles*, 179–220
[4] Lim E K, Kim T, Paik S, Haam S, Huh Y M and Lee K 2015 *Rev* **115** 327–94
[5] N. Azizi, N. Ladhari, M. Majdoub 2011 *Asian Journal of Textile* 1 130-37
[6] Polenz I, Brosseau Q and Baret J C 2015 *Soft Matter* **11** 2916-23
[7] Zhang H, Tumarkin E, Sullan R M A, Walker G C and Kumacheva E 2007 *Macromol. Rapid Commun.* **28** 527 – 38
[8] Damiati S, Kompella U B, Damiati S A and Kodzius R 2018 *Genes* **16**
[9] Shah R K, Kim J -W, Agresti J J, Weitz D A and Chu L -Y, 2008 *Soft Matter* **4** 2303-09
[10] Ma Y, Jiang Y, Tan H, Zhang Y and Gu J 2017 *Polym.* **9** 274.
[11] Gerven V T, Moors J, Dutre V and Vandecasteele C 2004 *Cem Concr Res.* **34** 1103-09 (2004)
[12] Quevedo E, Steinbacher J and McQuade D T 2005 *J. Am. Chem. Soc*, **127** 10498-499

IOP Conf. Series: Materials Science and Engineering **520** (2019) 012008 doi:10.1088/1757-899X/520/1/012008

A New Precision Goniometer Mechanism for Direct Angle Measurement by a Rotary Encoder

Mete Anakli [1,a] **and Caner Turker**[2]

[1]Aselsan Inc. MGEO Division, Ankara, Turkey
[2]Aselsan Inc. MGEO Division, Ankara, Turkey

[a] manakli@aselsan.com.tr; Tel: +903128475300

Abstract. In order to perform optical tests of an optical device such as a camera, a collimator system is needed which is aligned to a precision two axes rotary system (gimbal) where the optical device (Unit Under Test, UUT) is mounted. By the help of gimbal system, precision positioning of the UUT can be achieved and optical tests can be performed with a high accuracy. However, it is hard to position elevation axis accurately for relatively heavier UUT because of direct moment applied by UUT to the elevation axis of the gimbal system. Therefore, directly driven elevation axis is replaced by a goniometer system. Due to gear coupling clearance in goniometer systems, precision positioning is not achievable. In order to overcome this problem, a new mechanism, which is driven by a power screw, is used for the goniometer and virtual center of rotation is moved within the goniometer for direct angle measurement by a rotary encoder. Validation test is performed by autocollimator and precise mirror calibrators to check the accuracy of the new goniometer system.

1. Optical Tests of Vision Systems

These tests can be summarized as below and accuracy of the test system must be at most 1/5 of the test tolerances.

- FOV (Field of View)
- FOR (Field of Range, range test of the vision system own gimbal)
- MRTD (Minimum Resolvable Temperature Difference)
- Parallax
- Boresight
- Laser Tests.

2. Description of the Full Test System

In order the perform optical tests which are defined in the previous section, a special test system is required. This special test system consists of a ground isolated chassis, collimator system and two axes gimbal. The Figure 1 below represents a typical test system.

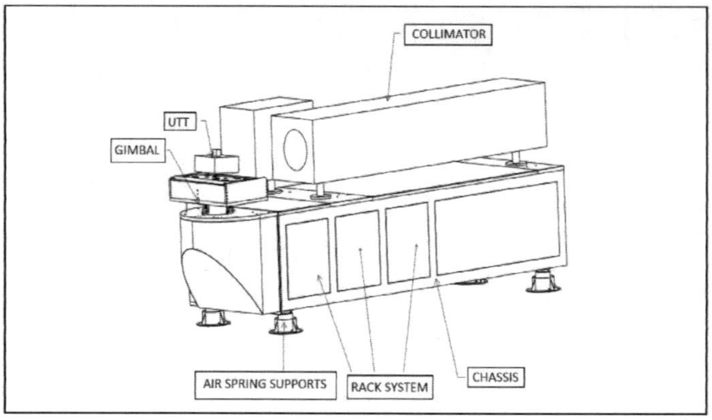

Figure 1. Full Test System

Chassis, which is the base of the full test system, has a number of air spring supports. These air spring supports are between ground and the chassis. They eliminate the ground low level vibration and they also maintain parallelism of the upper surface where collimator and gimbal system are mounted. All electronic modules (drivers, power supply, pc etc.) of collimator and gimbal system are located in the rack system.

Collimator system mainly consists of an optical target set and mirrors which simulate that target are placed at a specified test distance to UUT.

Gimbal system has two rotary axes which are called as azimuth and elevation. This fine electro-mechanical system is used for positioning of the UUT precisely according to test procedures. In this study elevation axes of precision gimbal types which can be used in optical tests will be investigated.

3. Optical Test Oriented Gimbal Elevation Axis Types

We can divide gimbal types into three. Advantages and disadvantages of each type will be mentioned in the related sections as below:

3.1. Traditional Gimbal System

This gimbal system has a wide use because of its working range. When test specifications need more than 90 degrees movement for elevation axis, this design is the best solution. Figure 2 shows a traditional gimbal system.

However, there are some disadvantages. Firstly, due to shape of the gimbal system, UUT mounts between gimbal columns and it could be hard to reach adjustment mechanisms of the UUT during tests. In addition, payload is high for relatively heavy UUTs, which causes motor stabilization problem for the elevation axis. Counter balance weight is not a complete solution because UUT weights vary. Another issue is the accuracy of the gimbal axes. High accuracy can be achieved by servo motors with no gearboxes to design a backlash free system which eliminates mechanical uncertainty if other mechanical elements of the system are fine as well. In this case, for high torque requirements, bigger motors are needed which result much expensive solutions.

On the other hand smaller direct drive motors can be used with fine gear boxes to meet torque requirement. In this case, encoder must be mounted on the output shaft.

Uncertainty of the system can be eliminated by active control of the motor according to encoder position.

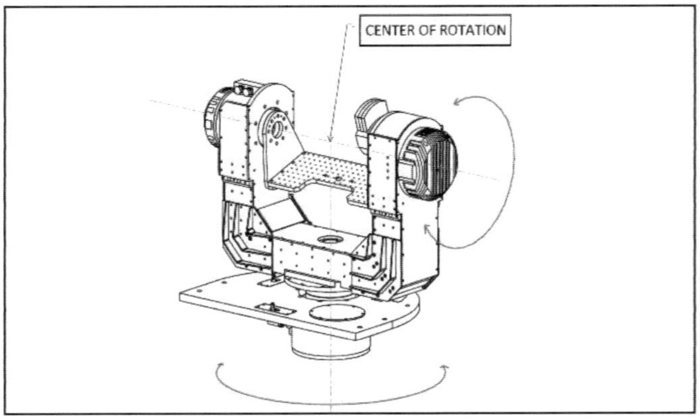

Figure 2. Traditional Gimbal System

3.2. Goniometer Systems

Goniometer systems provide an easy access to UUT adjustment points during tests because there are no column like structures to support elevation axis. Due to reduction ratio of the worm gear set, higher torques can be achievable by smaller motors for relatively higher UUTs. However, there is a clearance in the worm gear set in order to operate gear couple, which make the system less accurate. Backlash level of the gear set is increased due to metal abrasion while operating. In addition, angle of rotation of elevation axis is calculated by using rotation of the motor shaft which is highly affected by the system backlash level. A typical goniometer system can be seen in Figure 3.

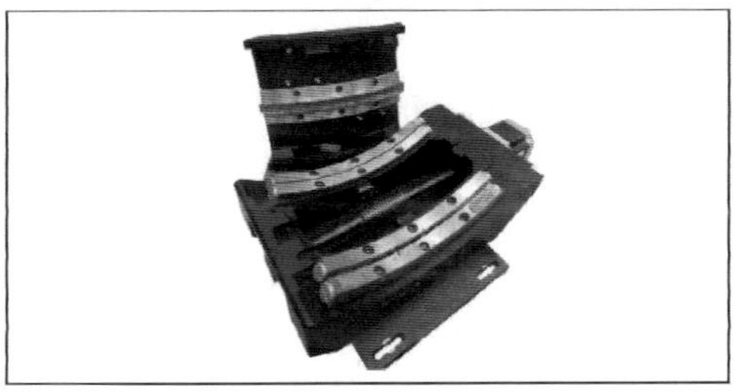

Figure 3. A Goniometer System

3.3. The New Goniometer System

This new mechanism also offers an easy access to UUT adjustment points during tests because there are no column like structures to support elevation axis. Due to usage of fine power screws for driving axis, higher torques can be achievable by smaller motors for relatively higher UUTs. In addition, backlash level is minimized as compared to worm gear set and metal abrasion effect is negligible due to bearing contact elements such as balls, cylinders etc.

The most important innovation for the new goniometer system is the moving of the goniometer rotation axis in the base of goniometer structure. Thus, there is no column like structures which can block to reach UUT adjustment points. On the other hand, by using the rotation axis in the base of goniometer structure, direct angle measurement can be obtained more accurately with a proper encoder mounted on the axis. As we see in Figure 4, we can derive that two angles α (goniometer rotation angle) and β (encoder rotation angle) are the same in magnitude but they have different directions. In addition, mechanical elements which are used to create the mechanism can be seen in Figure 5.

IOP Conf. Series: Materials Science and Engineering **520** (2019) 012008 doi:10.1088/1757-899X/520/1/012008

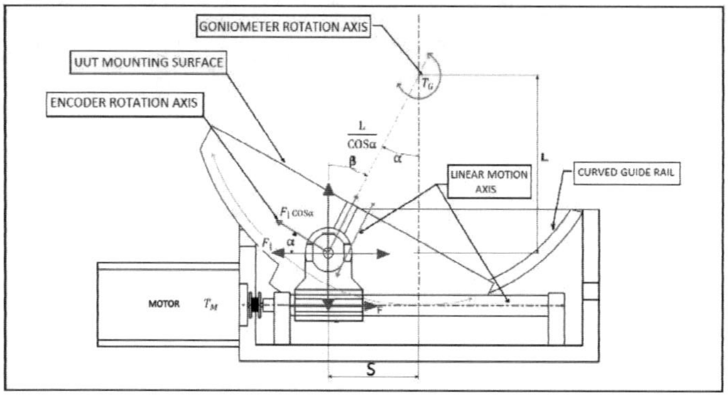

Figure 4. Rotation Angles with respect to Real and Virtual Rotation Axes and Force Distribution

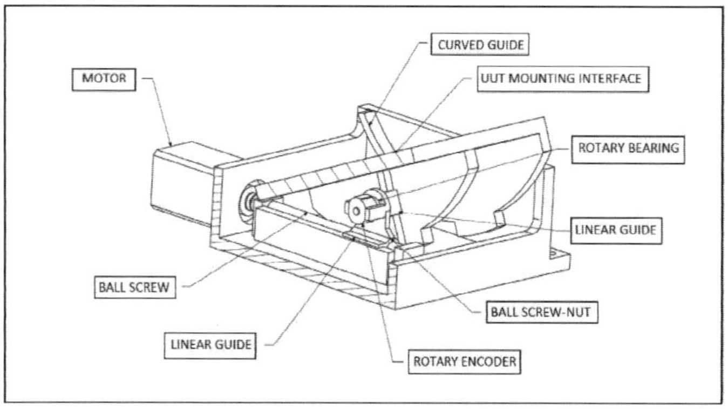

Figure 5. Mechanical Elements of the New Goniometer System

Accuracy level of the new goniometer system can be checked by a test setup which can be seen in Figure 6. This test setup consists of a theodolite [1] and a calibration mirror set [2] which is mounted to UUT mounting interface. The goniometer which was used in measurement process was designed for +/- 24 degrees due to design parameters. There are 3, 5, 15 degrees mirrors and a cubicle mirror in calibration mirror set and each has a 1 arc second accuracy. Measurement was taken by adjusting the objective of the theodolite as the cubicle mirror and the angular mirror which is stuck on the cubicle mirror can be seen clearly. When we stick an angular mirror on the normal surface of the cubicle mirror (i.e. the 3 degree mirror), we can turn the goniometer 3 degrees in order to make the angular mirror normal to the theodolite. Whether the center of theodolite is coincided with the center of the reflected light ray, we conclude the accuracy level of the goniometer system by using the angular change in theodolite. Measurements were taken for 3, 5, 15, 20 degrees in both +/- directions respectively and data is tabulated as in Table 1.

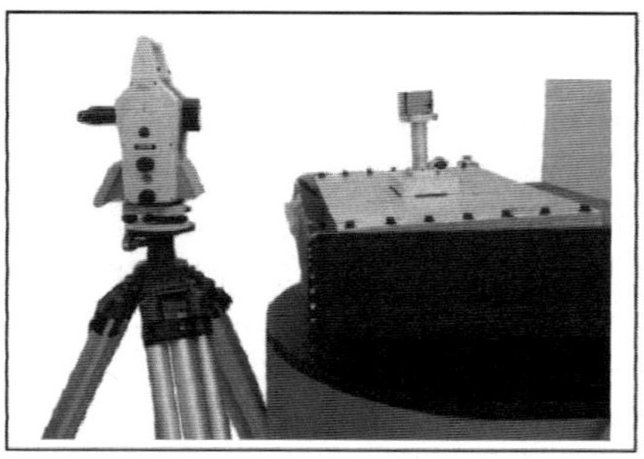

Figure 6. Test Setup

Table 1. Test Results

Calibration Mirror Angle (Degree)	Encoder(Degree)	Theodolite (Degree.Arc Minute. Arc Second)
0	0	90.00.24
-3	-3	90.00.24
-5	-5	90.00.24
-15	-15	90.00.40
-20	-20	90.00.24
0	0	90.00.24
3	3	90.00.24
5	5	90.00.26
15	15	90.00.35
20	20	90.00.38

Achievable torque level of the new system can be estimated by the help of Equation (1) below [3]. Power screw translation force and its moment arm with respect to goniometer rotation axis can be seen in Figure 4. If we substitute Equation (2) into Equation (3) below [4], by using a direct drive motor with 1 N.m output torque, 549 N.m torque can be achievable on new goniometer system. If we compare new goniometer system with a direct drive motor which produces 80 N.m torque, cost and mass of the compared motor are increased abruptly with respect to the one used in the new goniometer system.

T_m: Driving Torque [Nm]
F: Thrust Force [N]
r: Screw Lead [m]
μ: Efficiency
T_G: Goniometer Torque [Nm]
F_i: Inverse Thrust Force (On Encoder Center) [N]
L: Normal Moment Arm [m]
ω_m: Motor Angular Velocity [rad/s]
t: Time[s]
s: Stroke of Linear Displacement Power Screw Nut[m]
α: Goniometer Angle Displacement [rad]
ω_g: Goniometer Angular Velocity [rad/s]

$$T_m = \frac{F.r}{2\pi.\mu} \quad (1)$$

$$F_i = F = \frac{T_m.2\pi.\mu}{r} \quad (2)$$

$$T_G = \frac{F.cos\alpha.L}{cos\alpha} = F_i.L \quad (3)$$

$$T_G = F_i.L = \frac{T_m.2\pi.\mu}{r}.L \quad (4)$$

For new goniometer mechanism ;
L=0.230 m
r=0.002m
T_m=1 Nm
μ=0.95

$$T_G = F_i.L = \frac{T_m.2\pi.\mu}{r}.L=\frac{0,8.2\pi.0,95}{0,002}.0,23=549 \text{ N} \quad (5)$$

IOP Conf. Series: Materials Science and Engineering **520** (2019) 012008 doi:10.1088/1757-899X/520/1/012008

On the other hand, angular velocity equation of the goniometer system can be derived by using information in [5] as below;

$$\omega_m . t . r = s \tag{6}$$

$$tan(\alpha) = \frac{s}{L} \tag{7}$$

$$\alpha = invtan(\frac{s}{L}) = invtan(\frac{\omega_m . t . r}{L}) \tag{8}$$

$$\frac{\omega_m . \mathrm{r}}{L} = c \tag{9}$$

$$\alpha = \mathrm{invtan(c.t)} \tag{10}$$

If we integrate Equation (10) with respect to time by using a commercial math software, we can find angular velocity equation of the goniometer system as Equation (11).

$$\omega_g = \frac{d\alpha}{dt} = \frac{c}{c^2 t^2 + 1} \tag{11}$$

4. Conclusion

As a conclusion, this goniometer system can be designed up to +/- 90 degrees working range theoretically but size limits of the new goniometer system must be taken into consideration undoubtedly. Moreover, if we investigate the Table 1 again, accuracy of the goniometer system is nothing but the resolution of the encoder used [6]. In this system, resolution of the encoder was 16 bit which means 19,77 arc seconds readout uncertainty. By using an encoder with higher resolution, higher accuracy can be achievable.

In addition, one can conclude that for a constant ω_m , ω_g is reduced by time and maximum ω_g is acheived at home position (α=0).

L/r ratio can be considered as the reduction ratio of the system. Torque and velocity requirements can be adjusted by proper selection of L/r ratio.

References

[1] https://www.hexagonmi.com/products/industrial-theodolites-and-laser-stations/leica-tm6100a industrial-theodolite
[2] http://www.starrett.com/metrology/product-detail/AG%2016.%20R
[3] https://www.thk.com/sites/default/files/documents/uk_pdf/product/general/a/ee_A15.pdf
[4] Eres Söylemez, *Makina Teorisi-2 Makina Dinamiği*, (Birsen Yayınevi, İstanbul, 20073)
[5] R. C. Hibbeler, *Engineering Mechanics: Dynamics*, 12th Edition
[6] http://www.beisensors.com/pdfs/chm5-ssi-optical-absolute-encoder_en.pdf

IOP Conf. Series: Materials Science and Engineering **520** (2019) 012009 doi:10.1088/1757-899X/520/1/012009

Machine Learning for Tool Wear Classification in Milling Based on Force and Current Sensors

M Schwenzer[1,*], K Miura[1] and T Bergs[1]

[1] Laboratory for Machine Tools and Production Engineering (WZL) of the RWTH Aachen University, Campus-Boulevard 30, 52074 Aachen, Germany

*M.Schwenzer@wzl.rwth-aachen.de; +49-241-80-28021

Abstract. This paper presents how to classify the wear state of an end-mill based on force and current signals of the linear axes. The data is divided in binary classes based on the maximum flank wear. A support vector machine and a random forest are trained on orthogonal cutting experiments, but the validation is performed on arbitrary tool paths. To achieve this unique level of generality the signals are transformed into the rotation tool coordinate system. The features are extracted over five cutter revolutions. Support vector machines outperform random forests achieving 99,8% and 97% accuracy in the two classes on the test data and 98% and 61% accuracy on the validation data.

1. Introduction

Milling is a very flexible and economic production process, in particular suited for processing metals. Tool wear is as unavoidable as it is undesirable and influences manufacturing costs, workpiece quality and process safety (tool breakage). Nowadays, milling tools are replaced either after a fixed service life or based on the experience of the machine operator. Seldom, monitoring systems are used that trigger alarms if a fixed threshold is exceeded. This is only applicable for highly repetitive manufacturing tasks and the optimal choice of the threshold is a time-consuming task.

Only recently, data-driven models have turned into a hot topic in engineering with the rise of machine learning techniques, which have vastly spread throughout the academic community. They can learn abstract thresholds automatically and; thereby, enable for an automatic tool wear monitoring. However, most proposed classification models lack of generality.

This work trains models on data from orthogonal cutting but validates them on data from a complex tool path in one plane. To our known extend, this is the first work that dares to apply its model to structurally different data.

Though many works consider acoustic emission (AE) signals to classify tool wear [1–4], Kaya et al. [2] found that they only make a small contribution to an accurate classification of tool wear. In contrast to that, acceleration signals show a high correlation to all wear classes, whereas force is of particular interest to identify sharp tools. Kaya et al [2] use a genetic algorithm to optimize the selection of the features, which serve as input to a support vector machine. The selection could be made between 18 statistical values in time-domain of eight signals (force, acceleration, both in all Cartesian axes; torque, and AE). Analyzing the choice of the global optimization algorithm, they derived which signals have a high correlation to tool wear.

Nevertheless, the dominating input to models that classify tool wear is force, since its variation correlates to the wear of the tool [5]. Most approaches with machine learning techniques rely on force signals [3,4,6,7]. However, if the tool life duration is considered as input feature, it dominates the approximation [1]. This changes drastically if more than one tool is considered. Considering the tool life makes the approach prone to uncertainties in the production of the tool, the process, and the workpiece material.

The recent focus of learning tool wear lay on classification. Few studies approximate a tool wear curve by using a SVM for regression (SVR) [1,6]. However, a classification of the tool wear state was achieved afterwards via a customized threshold [6]. The current research focuses on support vector machines (SVMs) [1–3,8] and decision trees mostly random forest (RF) [3,4,9] as algorithms to classify the tool wear state.

For tool wear classification, popular features (signal describing characteristics) are the maximum, the mean, the median, and the standard deviation [2–4] of a signal. The choice of the features is key to the performance of every machine learning algorithm. Kaya et al. [2] extract less common features, such as the skewness or the kurtosis, and use a genetic algorithm to select the most appropriate features. On this background, the approach of Martínez-Arellano et al. [7] is exceptional since it does not extract features explicitly but uses deep learning on the raw signals. They transform the time-series signal to an image and apply a convolutional neural network (cNN) to classify the tool wear state. The approach considers the time-dependency of the force signal, making it unique for every cutting operation. This results in a lack of transferability to a new operation. However, the low level of generality is the major drawback of all works in this field. All studies in literature perform straight cuts for both, training and testing. None of the models was validated on structurally different data.

The described works present an active field with a strong focus on feasibility. No consolidation has yet taken place with regard to a method or certain features. Furthermore, most works lack of generality since training and testing data were taken from the same experiments. This work tackles the latter deficit and gathering the most promising features and algorithms from literature.

2. Approach

This work uses two of the most popular algorithms for classification from the field of machine learning [10]: Support vector machines (SVMs) and random forest (RF), which are basically decision graphs. The algorithms are particularly suited for problems with biased numbers of examples, i.e. many examples of one class (e.g. "good tool") and only a few examples for a different class (e.g. "dull tool").

2.1. Support vector machine

Originally designed for classification [11,12], SVMs have turned into a powerful tool to approximate arbitrary functions with a comparably small number of model parameters. This makes them less prone to the problem of "overfitting", a phenomenon that all supervised learning methods suffer from. It is best described by memorizing the training data revealing a lack of generalization. Models with a large number of parameters likely learn a training set without error but exhibit poor generalization, which makes the SVM of particular interest [11] in contrast to e.g. artificial neural networks (NNs) [13]. Today, it is among the most popular algorithms in machine learning because of its robustness and generalization.

SVMs determine the optimal hyperplane

$$\boldsymbol{w} \times \boldsymbol{x}_i + b = 0 \quad \text{with} \quad \boldsymbol{w} \in \mathbb{R}^n, b \in \mathbb{R} \tag{1}$$

that separates two groups of n data points $\boldsymbol{x}_i$ (were × denotes the dot product in the feature space). Assuming a set of data points with corresponding labels of

$$\{(\boldsymbol{x}_i, y_i) \mid i = 1, \dots, n; \boldsymbol{x}_i \in \mathbb{R}^n; y_i \in \{-1, +1\}\}, \tag{2}$$

an optimal hyperplane has to satisfy the condition

$$y_i = \operatorname{sgn}(\boldsymbol{w} \times \boldsymbol{x}_i + b). \tag{3}$$

The optimal hyperplane maximizes the minimum distance (often called "margin") of all data points to the hyperplane. This corresponds to minimizing the Euclidian norm of the normal vector w of the hyperplane. Those points, which lie on the margin – and thereby define the hyperplane – are the support vectors.

Boser et al. [11] point out that though a maximum margin optimization minimizes the probability of misclassification, it is prone to "atypical" data near the decision boundary. They argue that those data points have a very high influence on the decision boundary (through their high Lagrangian multiplier). This is counteracted by introducing so-called "slack variables"; variables that penalizes the extend of misclassification (i.e. the distance to the decision boundary) in the optimization function. In this way, a certain misclassification is allowed if it contributes to the overall accuracy and; thereby, allows for a more generic decision boundary.

2.2. Random forest

The RF consists out of multiple decision trees, of which everyone is build on a randomly selected subset of the feature space [14]. The RF obtains a decision by following the majority vote of its decision trees.

The algorithm does not use any pruning techniques on the decision trees. [14] argues that by pruning back decision trees they lose their accuracy on the training data, which gives little confidence of performing good on unseen data. The hope for generalization is bought at the cost of a lower accuracy on the training data. Instead, overfitting of a single classifier can be compensated by using multiple classifiers. This work builds a RF with 400 binary decision trees.

2.3. Experimental set-up

The experiments were conducted on a Mazak i600 VARIAXIS machining center with a Siemens SINUMERIK 840D sl controller. The tool was a two-fluted solid carbide end mill (diameter D = 10 mm, helix angle $\beta = 45$ °) and the material a high-alloyed steel (X5CrNi-18-10). In order to provoke tool wear, the process parameters were chosen somewhat more aggressive than suggested by the tool manufacturer: cutting speed v_c = 140 m/min, feed per tooth f_z = 0.1 mm. This results in a tool life travel path of 88 m, which consists mainly from straight cuts (550 times) and two simple impeller geometries with ten blades each, Figure 1. The idealized impeller geometry was chosen in order to represent a typical freeform surface geometry for milling and to emphasize the industrial applicability of this approach. It evaluates the transferability to a 2D cutting process.

The first impeller was manufactured after 350 cuts with a still sharp tool and the second impeller at the end with a dull tool. The tool state was checked at regular intervals on a microscope (Keyence VHX 5000). Those measures classify the tool whether it is sharp or dull. In this way the last 50 straight cuts and the last ten blade cuts were classified as "dull".

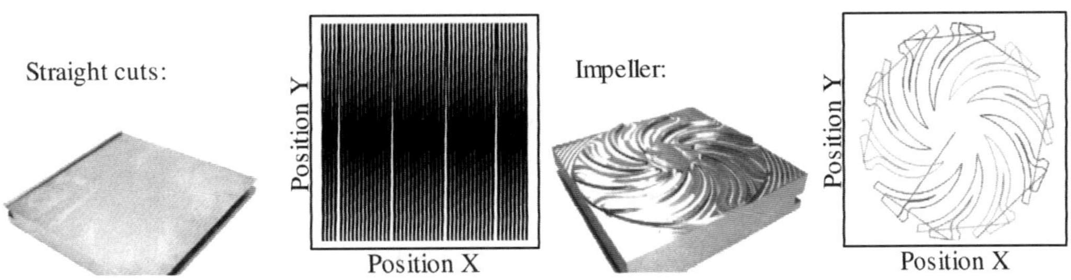

Figure 1. Overview of the two workpiece geometries: straight cuts and impeller.

The three-phase alternating current and voltage signals were measured at fs = 10,000 Hz with external sensors for the linear axes. To calculate the corresponding DC-components, the angular position of the motor was observed applying the method of [15]. In addition, the force was measured with a piezo-electric dynamometer (Kistler 9255) at fs = 10,000 Hz. Internal signals from the numeric controller (NC) of the machine were acquired via the compile cycle "axis data stream" (ADAS). This enables us to record the position, velocity, acceleration, and current information of every axis. The compile cycle works at fs = 400 Hz.

2.4. Preprocessing

2.4.1. Transformation To generalize from orthogonal cutting data to arbitrary conditions, the signals are transformed to the coordinate systems of the cutting edges, i.e. being split into their tangential, radial, and passive components. For this it is necessary to know the feed direction angle α and the angle of every cutting edge, Figure 2. Since the positions of the cutting edges do not vary to each other – unless a severe failure of the cutting edge occurs – its position angle composes of the rotation angle $\varphi(t)$ and the starting position angle $\Phi_{0,z}$.

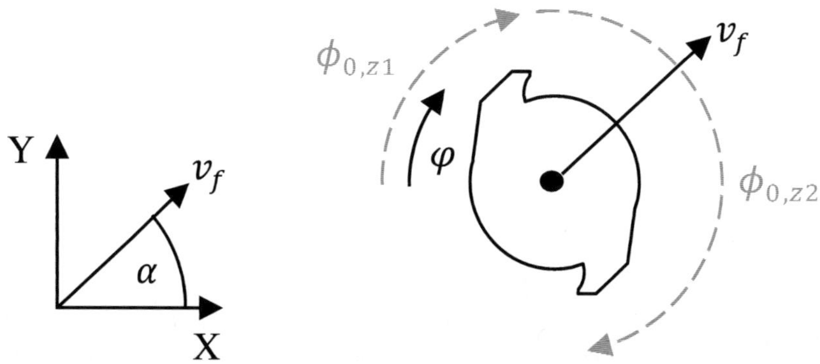

Figure 2. Schematic overview of the geometric values for the transformation from machine to tool coordinate system.

In general, both variables are unknown. While the starting position of the cutting edges must be identified through precise and laborious measurements, the angular position of the spindle is a vital information for the cascaded controllers of the spindle itself. However, modern machining centers work on encrypted BUS systems denying access to this information. Only external encoders with an open interface can provide the exact information. A practical approach to overcome both issues is to reconstruct the angular position of the cutting edges. We performed a cross correlation every ten cutter revolutions, restricting the allowed lag to $360/N_z$ of a revolution, where N_z is the number of teeth. This ensures to maintain the correct mapping between signal fraction and cutting edge.

Since the sampling rate is not a multiple of the rotational speed, the rotation angle $\varphi(t)$ varies and with it the uncut chip thickness h. This ensures that we can generalize from constant width of cut ae to other width of cuts via the uncut chip thickness.

2.4.2. Features Force is among the most popular signals to use to classify tool wear [1–4,6,7]. Furthermore, acceleration shows a high correlation to tool wear [1,2,4,6]. Cho et al. [6] showed that an additional electrical information (power) improves the classification accuracy by 20 %. García-Nieto et al. [1] demonstrate the correlation of the current to the tool wear state. Though they focused on the spindle signals, we chose to use the axis current as additional signals of all linear drives. The NC provides additionally the signals of the spindle.

The most common mechanistic force models in cutting technology – such as the exponential Kienzle model [16] or the linear Altintas model [17] – rely on the uncut chip thickness h and the uncut chip width b. This information is of particular importance to generate a generic classification since it scales the cutting force and all dependent signals. The uncut chip thickness h and the uncut chip width can be calculated through a simulation of the engagement conditions.

Table 1 provides the available signals of the experimental set-up.

Table 1. Overview of the available signals and calculated additional data.

External signals ($f_s = 10\,kHz$)	Force (F_t, F_r, F_p), current ($I_{q,t}$, $I_{q,r}$, $I_{q,p}$), uncut chip thickness h, uncut chip width b
Internal signals ($f_s = 400\,Hz$)	Torque (T_r, T_t, T_p, T_S), current ($I_{q,t}$, $I_{q,r}$, $I_{q,p}$, $I_{q,S}$), acceleration (a_t, a_r, a_p), uncut chip thickness h, uncut chip thickness b

The mean, the maximum, and the standard deviation were extracted over five cutter revolutions building the features space. Most of the literature uses those statistical values from time domain as features (see [2–4]). The accuracy of the two machine learning algorithms is tested on three different feature sets consisting out of the same data points:

- features extracted from internal signals,
- features extracted from external signals, and
- merging the previous data sets.

2.4.3. Data division The major drawback of all classification models proposed in literature is its lack of generality. Trained and tested on the same cutting geometry (often simply straight cuts), the model cannot be transferred to other workpiece geometries. To prove the ability to generalize of our approach, we train the model on straight cuts but validate it on the data from the idealized impeller, Figure 1, representing structurally different data. We use 70 % of the data from the straight cuts to train the classification model; the remaining 30 % is used for testing. Table 2 provides the absolute number of data points.

Table 2. Overview of the different data sets.

	Training	**Testing**	**Validation**
Total number of data points	65553 (Straight cuts)	28094 (Straight cuts)	6023 (Impeller)
Number of "sharp" data points – class A	59667	25572	3012
Number of "dull" data points – class B	5886	2522	3011

3. Results and discussion

3.1. Results

To evaluate the classification accuracy of the two machine learning algorithms on the different feature sets, Table 3 and Table 4 provide the confusion matrices of the test data and the validation data respectively. The row of a confusion matrix represents the estimated classes (indicated by the accent "^") and the columns represent the actual classes (class A: "sharp", class B: "dull"). A high value on the diagonal axis indicates a good performance. For the sake of convenience, we colored the cells of the matrix in traffic light colors. The secondary diagonal contains the number of wrong assignments.

Table 3. Results of the test data (straight cuts) for different feature sets (only signals from internal or external sensors and joint performance) and both algorithms: support vector machine (SVM) and random forest (RF). Class A: "sharp tool", class B: "dull tool".

Feature set / Algorithm		Intern A	Intern B	Extern A	Extern B	Intern + Extern A	Intern + Extern B
SVM	$\hat{A}$	100 %	100 %	99,8 %	2,7 %	99,9 %	1,3 %
	$\hat{B}$	0 %	0 %	0,2 %	97,3 %	0,07 %	98,7 %
RF	$\hat{A}$	98 %	21 %	99,9 %	2,5 %	99,9 %	1,2 %
	$\hat{B}$	2 %	79 %	0,1 %	97,5 %	0,03 %	98,8 %

Table 4. Results of the validation data (idealized impeller geometry) for different feature sets (only signals from internal or external sensors and joint performance) and both algorithms: support vector machine (SVM) and random forest (RF). Class A: "sharp tool", class B: "dull tool".

Feature set / Algorithm		Intern A	Intern B	Extern A	Extern B	Intern + Extern A	Intern + Extern B
SVM	$\hat{A}$	100 %	100 %	98 %	39 %	100 %	42 %
	$\hat{B}$	0 %	0 %	2 %	61 %	0 %	58 %
RF	$\hat{A}$	98,2 %	99 %	87 %	69 %	99 %	99 %
	$\hat{B}$	1,8 %	1 %	13 %	31 %	1 %	1 %

3.1.1 Feature sets Regarding the feature sets, the internal signals alone are not suitable for tool wear classification. The random forest is able to achieve a solid classification accuracy on the test data, which has the same structure as the data on which the algorithm is trained. However, it is unable to generalize from it (only 1 % accuracy on class B, "dull", Table 4). Using the features extracted from external signals results in an outstanding accuracy on the test data and even a solid accuracy on the validation data. This is in good accordance with the literature – which only tests on data that has the same structure as the training data. The accuracy and the sensitivity of the external sensors (dynamometer, current, and voltage sensors) and the sampling frequency is much higher than the internal signals provided by the machine controller. This drastically affects the accuracy of the classification algorithms.

At the first glance, it is surprising that merging the features of the internal and the external signals does not lead the same or a higher classification accuracy as the features of the external signals alone. A possible reason for this is that the features extracted from the internal signals do not contain suitable information (bad accuracy if they are used alone) and increase the dimension of the feature space to

19 · 3 = 57 dimensions. The problem of a reduced significance of the individual feature in a high-dimensional feature space is known as "curse of dimensionality". The feature space of the external signals has 6 · 3 = 18 dimensions; the feature space of the internal signals 13 · 3 = 39 dimensions.

3.1.2. Classification algorithms Both, the SVM and the RF, achieve a very high accuracy on the test data. Therefore, for evaluation purposes, we focus on the validation data, Table 4. The way in which randomization is introduced in multiple decision trees – by using a random subset of the feature space – makes RF particularly prone to the curse of dimensionality. If the majority of the features does not contribute to the classification objective, many trees are grown that have a low classification accuracy. Since RF relies on the majority vote of these trees, this severely affects the classification accuracy of the forest. Using more classes eases this problem as the wrong classifications may distribute among the more classes and a lower number of correct trees is required for a majority vote. Solving an optimization problem, the SVM is less prone to the curse of dimensionality.

3.2. Limitations and discussion

Examining the points that are classified incorrectly more closer, Figure 3, reveals that both algorithms manage to classify sharp tools (class A) correctly: average accuracy 98.1 % (SVM) and 86.1 % (RF). Figure 3 depicts the accuracy rate in groups of 120 samples of the validation data. Note that the first 3012 samples represent a sharp tool, whereas the remaining 3011 samples represent a dull tool.

Both algorithms experience a severe drop in accuracy when the tool wear class changes to "dull" (class B). However, as the tool wear increases the SVM is able to identify the dull tool correctly: average accuracy of the last six blades 99.2 % (SVM) and 18.9 % (RF).

The arbitrary definition of a "sharp" and a "dull" tool may cause the problem of an ambiguous separation plane. Tool wear evolves continuously, which makes it difficult to define the exact point from where a tool is considered as "dull". The sensor signals reflect this problem as they change gradually. Only if the tool suffers from breakouts, one may obtain a distinct separation plane.

The presented system itself suffers from the inaccuracy introduced by the missing rotational position of the spindle (which was reconstructed through cross correlation) and of the auxiliary drives (which was estimated through a Kalman filter). To exploit the full potential of this approach, the encoder signal of all motors must be known. In future research, the performance over multiple tools must be examined; as well as the sensitivity to more classes.

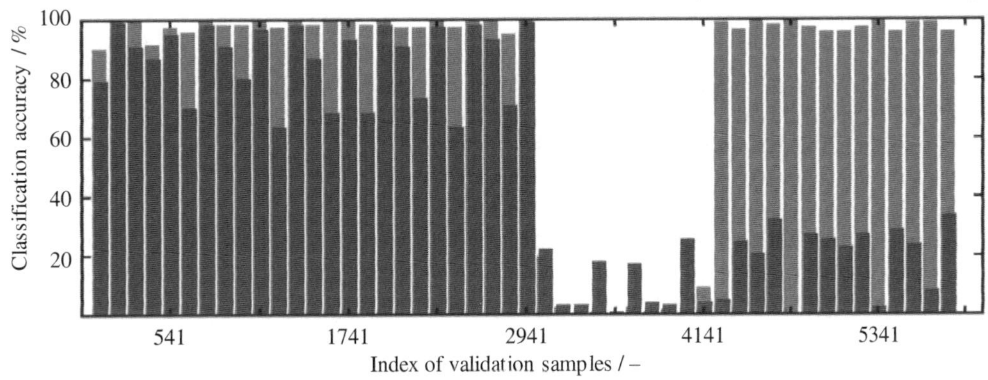

Figure 3. Evolution of the classification accuracy in the validation data. Blue bars: SVM, gray bars: RF

3.3. Notes to a practical application

Within the rising digitalization and activities such as "Industry 4.0" or an "Internet of Production", the production process itself becomes transparent. Machine tool manufacturers open their interfaces and include more sensors into their products. This work stresses the significance of signal quality but is based on standard measurement equipment. Machine learning in particular provides a method to automatically analyze big data and robustly monitor e.g. tool wear. Considering the engagement conditions provided by some CAM programs and transforming all signals to the coordinate system of the tool enables to learn a model that is independent of the tool path and thus is even suited for single-batch-production. All preprocessing and training can be conducted onsite on an edge computer – the resulting machine learning model may even be transferred to other machine tools.

4. Conclusion

This paper presents the first approach on tool wear classification that is validated on structural different data. This previously unreached level of generality is achieved by transforming the signals from the fixed machine coordinate system to the rotating coordinate system of the cutting edges. It is the first to use classical process characteristics such as the uncut chip thickness h and the uncut chip width b. Features are extracted from signal segments of five cutter revolutions, contributing to the objective of creating a model as generic as possible. Furthermore, this results in more data points than any study in literature.

Two popular classification algorithms were examined: SVM and RF. Both achieve a very high accuracy on the test data (over 97 % for both classes). But SVM is able to obtain a more generic model from the straight cutting tests: 98 % for class A ("sharp") and 61 % for class B ("dull") on the blade geometry.

Examining features from internal sensors (torque and current at fs = 400 Hz) and external sensors (force and current at fs = 10 kHz) indicates that

- the higher quality and sampling rate of external sensors lead to a much more accurate classification; and that
- adding the features extracted from the internal sensor worsens the performance ("curse of dimensionality").

With the raise of the digitalization at shop floor level more data and of higher quality will become available. This work outlines an automatic analysis and lies the basis to profit from data.

5. Acknowledgments

The authors would like to thank the German Research Foundation DFG for the kind support within the Cluster of Excellence "Internet of Production" (Project ID: 390621612).

References

[1] García-Nieto P J, García-Gonzalo E, Vilán Vilán J A and Segade Robleda A 2016 A new predictive model based on the PSO-optimized support vector machine approach for predicting the milling tool wear from milling runs experimental data *The International Journal of Advanced Manufacturing Technology* **86** 769–80

[2] Kaya B, Oysu C, Ertunc H M and Ocak H 2012 A support vector machine-based online tool condition monitoring for milling using sensor fusion and a genetic algorithm *Proceedings of the Institution of Mechanical Engineers, Part B: Journal of Engineering Manufacture* **226** 1808–18

[3] Wu D, Jennings C, Terpenny J, Gao R X and Kumara S 2017 A Comparative Study on Machine Learning Algorithms for Smart Manufacturing: Tool Wear Prediction Using Random Forests *Journal of Manufacturing Science and Engineering* **139** 071018

[4] Wu D, Jennings C, Terpenny J and Kumara S 2016 Cloud-based machine learning for predictive analytics: Tool wear prediction in milling *2016 IEEE International Conference on Big Data (Big Data)* 2016 IEEE International Conference on Big Data (Big Data) (Washington DC,USA: IEEE) pp 2062–9

[5] Dimla D E 2000 Sensor signals for tool-wear monitoring in metal cutting operations—a review of methods *International Journal of Machine Tools and Manufacture* **40** 1073–1098

[6] Cho S, Asfour S, Onar A and Kaundinya N 2005 Tool breakage detection using support vector machine learning in a milling process *International Journal of Machine Tools and Manufacture* **45** 241–9

[7] Martínez-Arellano G, Terrazas G, Benardos P and Ratchev S 2018 In-process Tool Wear Prediction System Based on Machine Learning Techniques and Force Analysis 8th CIRP Conference on High Performance Cutting (Budapest, Hungary: Elsevier)

[8] Nieto P J G, García–Gonzalo E, Vilán J A V and Robleda A S 2015 Modeling the milling tool wear by using an evolutionary SVM–based model from milling runs experimental data International Conference on Computational Methods in Sciences and Engineering 2015 (ICCMSE 2015) (Athens, Greece) p 190021

[9] Wu D, Jennings C, Terpenny J, Gao R and Kumara S 2017 Data-Driven Prognostics Using Random Forests: Prediction of Tool Wear *Volume 3: Manufacturing Equipment and Systems* ASME 2017 12th International Manufacturing Science and Engineering Conference collocated with the JSME/ASME 2017 6th International Conference on Materials and Processing (Los Angeles, California, USA: ASME) p V003T04A048

[10] Wu X, Kumar V, Ross Quinlan J, Ghosh J, Yang Q, Motoda H, McLachlan G J, Ng A, Liu B, Yu P S, Zhou Z-H, Steinbach M, Hand D J and Steinberg D 2008 Top 10 algorithms in data mining *Knowledge and Information Systems* **14** 1–37

[11] Boser B E, Guyon I and Vapnik V 1992 A training algorithm for optimal margin classifiers *Proceedings of the 5th annual ACM workshop on computational learning theory* (ACM Press) pp 144–152

[12] Vapnik V, Golowich S E and Smola A J 1996 Support Vector Method for Function Approximation, Regression Estimation and Signal Processing *NIPS*

[13] Smola A J and Schölkopf B 2003 A Tutorial on Support Vector Regression

[14] Tin Kam Ho 1995 Random decision forests *Proceedings of 3rd International Conference on Document Analysis and Recognition* 3rd International Conference on Document Analysis and Recognition vol 1 (Montreal, Que., Canada: IEEE Comput. Soc. Press) pp 278–82

[15] Miura K, Döbbeler B and Klocke F 2018 Cutting Power Estimation via External Voltage and Current Sensors on Feed-Drive Axis for the Straight Milling Process *Procedia CIRP* CIRPe 2018- 6th CIRP Global Web Conferene

[16] Kienzle O 1952 Die Bestimmung von Kräften und Leistungen an spanenden Werkzeugmaschinen *VDI-Z* VDI-Z 81. VDI-Hauptversammlung der Fachgruppe Betriebstechnik vol 96 (VDI-Verlag) pp 299–305

[17] Altintas Y and Lee P 1996 A General Mechanics and Dynamics Model for Helical End Mills *CIRP Annals - Manufacturing Technology* **45** 59–64

Influence of Selected Parameters of the Motor Controller on the Current Characteristics of the DC Brush Motor Used in the Silesian Greenpower's Vehicle

Paweł Żur[1,a] , Andrzej Baier[1] and Alicja Kołodziej[1]

[1]Silesian University of Technology, Faculty of Mechanical Engineering, Konarskiego 18a, 44-100 Gliwice, Poland

[a] Corresponding author : pawelzur@gmail.com

Abstract. The paper presents tests of the controller created by Silesian Greenpower team for a DC motor carried out on an motor test stand. The main scope of the work was to check how selected parameters of settings affect current characteristics and speed of the motor in order to optimize the efficiency of the chain transmission. Charts of current and revolutions of the Fracmo PM62 motor, a description of the applied testing methodology, photos of the testing process and results of measurements carried out on the test stand have been described in the article. The research has resulted in a comparison of obtained outcomes, summary and presentation of conclusions, which concern practical remarks related to the further development of the control system applied in the Silesian Greenpower's car.

1. Introduction

The main scope of the research was to determine optimal current characteristics for Fracmo PM62 electric motor. This motor is powered by two 12 V batteries with a capacity of 36 Ah each and is used for driving the electric car created by Silesian Greenpower team. [1]

Regarding the fact that the car takes part in racing series, the priority was to optimize energy consumption at the start and during the race. [2]

Considering that motor controller used in research has never been tested on dyno station before, the original settings of its creator were tested at the very beginning. Next stages of the project were such adjustments that the energy consumption of the motor was optimized.

All tests have been carried out on the actual team's car – "Bullet II" in which chain transmission, Shimano derailleur, Shimano cassette and DT hub have been applied. Due to car's construction, only 3 out of 11 gears have been used in tests.

Interfaculty students' project Silesian Greenpower created at the Silesian University of Technology has been founded in 2010. Since then, the main goal of annual editions is designing and constructing an ecologic electric vehicle, which will take part in the Greenpower racing series. Besides competition, the project's aim is to promote ecological energy sources, especially in the automotive industry. [1]

Throughout the race, the goal is to pass as many laps as possible in a given time.

In the last season, the team has won such awards as Siemens Award for telemetry application, IET Award for innovations applied in the car and 4th place in F24+ league (Fig. 1).

Figure 1. Silesian Greenpower team and their trophies

One of the key tasks during designing an ecologic electric vehicle is minimizing the drag during a race and maximizing the efficiency of the drive system. Hence the need to ensure optimal motor operation through an efficient control system.

2. Test objective

2.1 Test stand

The inertial part of the motor test stand of the Silesian Greenpower team was used for the tests, along with the "Bullet II" car (Fig.2). In order to make current measurements, motor revolutions and further processing of obtained results, the "SG-Telemetry" measuring system by Adam Stalica was used.

Figure 2. Motor test stand along with the "Bullet II" car

2.2 Selected controller - "SG MotorController"

DC motor control is based on creating the appropriate voltage added to the commutator's brushes. The PWM (Pulse-Width Modulation) signal controls transistors – MOSFET in this particular case. Motor speed increases with increasing PWM signal. Examples of signal waveforms due to the PWM value are shown in Fig. 3. [3-6]

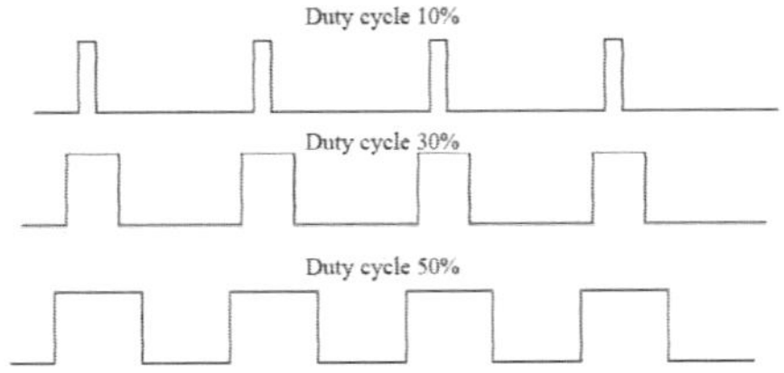

Figure 3. PWM duty cycle examples [7]

Selected controller, SG MotorController" is created by one of the Silesian Greenpower team members - Adam Stalica. In comparison to other available controllers on the market, it is possible to configure many parameters to optimize power consumption and ensure the best possible motor performance. Specification of "SG MotorController" has been presented in Table 1.

Table 1. Specification of "SG MotorController"

Parameter	Unit	Range
Normal voltage range	[V]	0-48
Minimum/maximum voltage	[V]	0-60
Motor current (max)	[A]	300
Motor current (1 minute)	[A]	250
Motor current (continuous)	[A]	200
Input signal	[V]	15-30
Input power	[V]	5-36
Reverse polarity protection	-	N
Ramping	[min]	0-4000
Switching frequency	[kHz]	0.001-5
Resistant (conduction)	[mΩ]	1.15

This controller enables integration with the aforementioned measuring system, which gives the possibility to control and change parameters in real time during the race. Parameters, which can be adjusted with its description have been presented in Table 2.

Table 2. Adjustable parameters of "SG MotorController"

Parameter	Range	Description
PWM Start	0-4000	Represents initial PWM
PWM UP-1	1-500	Value added to previous one in each cycle of ramp no. 1
PWM Max	1-4000	Maximum selected value of PWM
PWM UP-2	1-500	Value added to previous one in each cycle of ramp no. 2
PWM Fast Time-1	1-60 000	Time in which controller does not increase PWM in ramp no. 1
PWM Fast Time-2	1-60 000	Time in which controller does not increase PWM in ramp no. 2

Selected controller – "SG MotorController" allows adjusting two increasing ramps which, depending on selected parameters, set the PWM signal in various ways.

2.3 Test stages

The test consisted of three stages. The first stage was to inspect default values of the controller. The second stage – inspecting the same values with the release of the throttle for five seconds. The third stage of the research was manipulating parameters set in stage 2.

2.3.1 Inspecting default values of the controller Default values of the controller have been tested (Table 3.) and obtained results have been presented in Figure 4. In addition, during the tests, the difference in voltage was measured between this that was directly on the batteries and this that was delivered to the engine. The voltage summary, after the controller has reached the full PWM signal, is shown in Table 4.

Table 3. Default values of "SG MotorController"

Parameter	Value
PWM Start	800
PWM UP-2	15
PWM Max	4000
PWM Up-1	500
PWM Fast Time-2	1000
PWM Fast Time-1	150

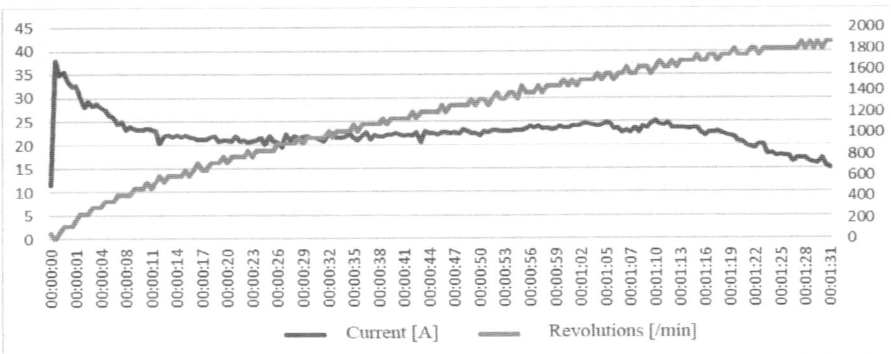

Figure 4. Motor characteristics with default values of the controller – derailleur set to 4th gear

Table 4. Voltage measured on batteries and the motor

Voltage	Value
Batteries	23.87 [V]
Motor	24.77 [V]
Difference	0.1 [V]

Values shown above mean that voltage loss is slight, almost marginal. For comparison, the same measurement was carried out under the same conditions on a commercial Kelly Controls KDS24100E controller and the difference was 0.78 [V].

IOP Conf. Series: Materials Science and Engineering **520** (2019) 012010 doi:10.1088/1757-899X/520/1/012010

2.3.2 Inspecting default values of the controller The test objective was to release the throttle by the driver which was supposed to simulate the braking of the car before entering the corner, and then re-engaging the gas, simulating the exit from the corner (Figure 5).

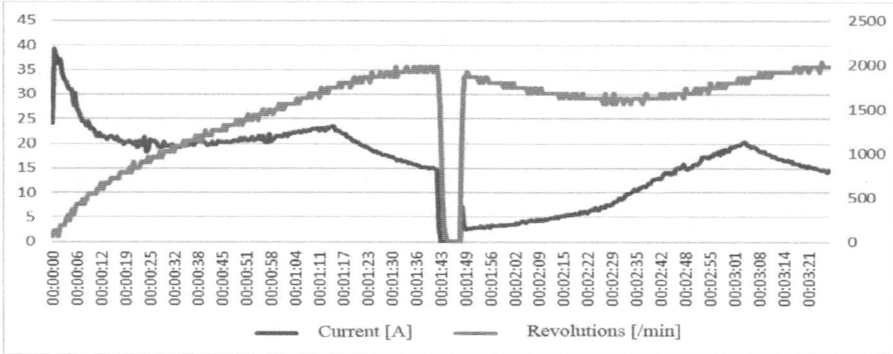

Figure 5. Motor characteristics with release of the throttle - derailleur set to 4th gear

It is worth to mention that drivers of the car usually do not use a mechanical brake (it causes unnecessary energy loss), and it is only sufficient to release the throttle.

2.3.3 Manipulating parameters of the controller The next step was modifying the parameters and observing the average current consumed by the engine. In the table below (Table 5) the results of the tests were collected.

Table 5. Results of the research

	Test 1	Test 2	Test 3	Test 4	Test 5	Test 6
Time to throttle release [s]	105	124	115	126	151	151
Average current [A]	27,32	24,85	25,44	25,81	23,88	24,14
Ramp 1 time [s]	24,26	24,26	20,60	20,60	20,60	20,60
Ramp 2 time [s]	2,5	9,3	9,3	13,8	27,4	20,6
Velocity of the car after ramp 2 [km/h]	56	52	54	52	48	50

As it can be seen above (Table 5), the ramp time has a significant influence on the average current consumed during the test. A short ramp results in a much higher current and a higher speed of the car. Obtained results should be used to set the controller for a specific track configuration.

3. Conclusions

1. Increasing "PWM UP_1" by 50 resulted in the sharpening of the starting ramp - the graph is linear with a greater inclination to the horizontal axis. As a result, the ramp has been shortened. Even though the current in the peak increased, the average current decreased significantly, which will have a positive impact on energy consumption during the race.
2. Fivefold reduction of "PWM Fast Time_2" caused in doubled time of the ramp. The speed of the car decreased and remained at a lower level for a longer time.

3. Setting the "PWM Fast Time_2" parameter to 200 resulted in ramp extension, but the average current of 24.14 [A] is higher than with parameter set to 100, and the time needed to stabilize the current has not changed.
4. Lowering the PWM UP_1 parameter below 150 caused malfunctions of the controller - the current was rising with a jumping manner creating steps on the graph.

The main purpose of the work was to test the new controller. Influence of individual parameters on characteristics of the Silesian Greenpower's motor has been shown. The research and development of the SG MotorController are being continued. In the future, the data collected will be used to compare the controller to commercial products and with the computer simulation prepared in Siemens NX software.

Acknowledgements
The paper has been financed from the funds of the project "Electric vehicle of Silesian Greenpower - student interfaculty project of the Silesian University of Technology" as part of a conceptual non-competition project entitled "Best of the best! 3.0 "Development co-financed by the European Social Fund under the Operational Program Knowledge, based on the application for co-financing POWR.03.03.00-00-P011 / 18.

References

[1] https://www.sg.polsl.pl [retrieved 28.11.2018]
[2] Baier, A., Zur, P., Kolodziej, A., Konopka, P., & Komander, M. (2018, August). *Studies on optimization of 3D-printed elements applied in Silesian Greenpower vehicle*. In IOP Conference Series: Materials Science and Engineering (Vol. 400, No. 2, p. **022010**). IOP Publishing
[3] https://ep.com.pl/artykuly/9121-Regulowanie_obrotow_napedow_elektrycznych.html [retrieved 28.11.2018]
[4] Urbanowicz H., Nowacki Z. 1986. *Napęd elektryczny w pytaniach i odpowiedziach*. WNT, Warszawa. ISBN 83-204-0762-1
[5] Kosow, I. L. (1964). Electric machinery and control. Prentice-Hall.
[6] National Fire Protection Association (2008). *Article 100 Definitions*. NFPA 70 National Electrical Code. 1 Batterymarch Park, Quincy, MA 02169: NFPA. p. **24**.
[7] Bakshi, U. A.; Godse, A. P. (2007). *§8.2 The depletion mode MOSFET*. Electronic Circuits. Technical Publications. pp. **8–2**. ISBN 978-81-8431-284-3. Bakshi, U. A.; Godse, A. P. (2007). *§8.2 The depletion mode MOSFET*. Electronic Circuits. Technical Publications. pp. **8–2**. ISBN 978-81-8431-284-3.

IOP Conf. Series: Materials Science and Engineering **520** (2019) 012011 doi:10.1088/1757-899X/520/1/012011

Acceptance Modelling in Product Development – Case Study: Connected Systems for Industry 4.0 Solutions

V Zimmermann [1,a], J Heimicke [1], A Albers [1] and N Reiß [1]

[1] IPEK – Institute of Product Engineering, Karlsruhe Institute of Technology (KIT), Kaiserstraße 10, Karlsruhe 76131, Germany

[a] Corresponding author: valentin.zimmermann@partner.kit.edu

Abstract. Within Industry 4.0 many connected systems are entering production to optimize the processes. The integration of the manual workstations into a Smart Factory, as a fully connected production, is required as well. Therefore, systems with which the production workers interact as users are introduced. However, as a side-effect, the production employees and their working performance are more transparent as well. This often leads to a lack of user acceptance, even though these systems have positive effects on complete production processes. If, for unforeseen reasons, there is a lack of user acceptance, the introduced system will not be as successful as planned, irrespective of how technically valuable the future product is. Therefore, the potential user acceptance towards the System-in-Development should be understood. This contribution introduces an acceptance model that has been developed and validated within a Live-Lab with industrial participation, as a research environment between laboratory and field studies. For this purpose, influencing factors for user acceptance of connected products were identified. Based on various scientific surveys, the factors were weighted and their interactions analysed in order to enable prognosis regarding user acceptance and to derive recommendations concerning actions for the downstream processes.

1. Motivation – Connected Systems for Industry 4.0 Solutions

In the course of Industry 4.0, the digital interlinking of all components transforms production into Smart Factories. [1] All value-adding networks are controlled in real time, where machines make partly or fully automatic decisions. [2] This is done on the basis of data previously generated, processed and stored along the value chain. This database provides increased transparency along the entire value chain and enables, among other things, the identification of potentials for optimization in the production process. [3] By connecting all components and generating data at the relevant workstations, it is possible to create a holistic digital image of the value chain, from order-entry to invoicing. This makes it possible to use production capacity more intelligent or to optimize individual manufacturing steps. [4] Recent studies show, that industry 4.0 solutions will increase the productivity of German machine builders' production facilities by an average of about 7-11%. [5] In addition, data acquisition in a production line also enables condition monitoring of all machines and thus ensures optimized maintenance cycles and therefore fewer breakdowns. [2] The achieved transparency also means that the work of the people, who are involved in the value-added process will also become more transparent. [6] It can be seen that the general attitude towards data collection, data storage and related issues is predominantly sceptical. [7] However, this is not only evident on a professional level, but it also plays a role in private life, in form of social networks such as facebook or google maps. In this respect, a preliminary study (Fig. 1) was carried out within the scope of this publication, in order to confirm the mood regarding data collection, which was identified in literature also in working life.

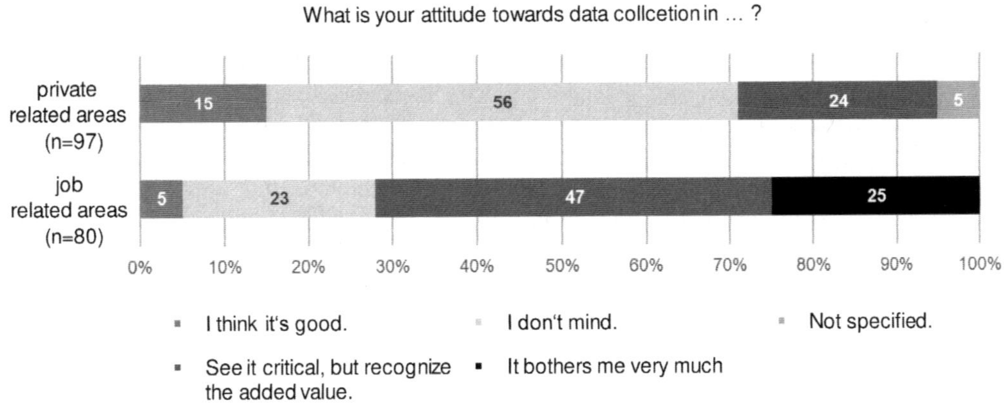

Figure 1. Attitude towards the topic of data collection in the private (at the top) and job-related (below) areas - own preliminary study

The results of the preliminary study show that social networks are widespread in private use, but that the majority of participants remains sceptical about data collection in their working environment. Most respondents state, they use such systems because they believe that their use will result in a personal advantage or there are no comparable alternatives. In the professional environment, scepticism predominates, especially since the personal benefit has not been recognized yet. For companies, which develop production systems, it is a necessary condition for their sustainable economic success that the production workers of their future customers accept and use these systems. This inevitably leads to the fact that the attitude of future users towards the product generation in development represents a decisive influence and success factor in the development process of these companies. Furthermore, globally operating companies cannot allow themselves to violate ethical values towards their employees and, for example, to implement systems that are not accepted by users. However, it is not possible to make a valid statement about the associated acceptance of potential users in the early stage of the development process, especially in the context of connected systems. For this purpose, in this contribution a methodology will be presented, by means of which the developer can model and compare the potential acceptance of different alternative solutions in the early stage of the development process.

2. State of the Art

2.1. Innovation and Diffusion of technical Products in the PGE – Product Generation Engineering

According to SCHUMPETER [8], an innovation differs from an invention by its economic impact. In the innovation process of product development, an idea that leads to a new product is technically implemented into an invention, which in turn has to be taken over by a broad mass on the market and thus penetrates the market (diffusion [9]). In order to achieve this goal, a company must use marketing activities and methods to achieve a suitable market launch of the product. [10] In order to be as successful as possible in this process, a product profile, which is a model-based description of the different benefits from the perspective of customers, users and the provider, forms the restrictive framework for generating ideas and inventions. These profiles describe the demand situation on the market, whilst taking into account the three above-mentioned perspectives and further elements like the competitive situation or the intended core functions of the future product. However, the product profile does not include the technical solution for the identified demand. [10] The profile is also used to continuously validate generated ideas and technical solutions; i.e. the invention, against the identified customer, user and provider benefits, thus increasing the probability of a future diffusion.

[11] The diffusion of an invention occurs through adoption by individuals, where adoption is the actual take-over. [12] At this point, the real customer, user and provider requirements are relevant. According to ALBERS [13], the development of new products follows the model of the PGE – Product Generation Engineering, which means the development of products in generations [13], is always based on reference products. These can be predecessor products, competitor products, or their subsystems. In addition, new generations are developed through the systematic combination of the activities Carryover Variation (CV), Embodiment Variation (EV) and Principle Variation (PV). [13] In the case of the CV, the embodiment design and solution principle of the subsystem remain unchanged compared with the reference product. This reduces the constructive change effort to a minimal and affect in particular the system boundaries. Within the EV, the existing principle of the reference product is retained, but the embodiment is adapted. In the course of the PV, subsystems are realized by implementing changed solution principles. [14]

The process of product development does not take place sequentially. Activities of product engineering are carried out iteratively and simultaneously over different phases and the engineer is supported by development methods. [15, 16] Within those processes the early and consistent integration of the customer (Co-Creation [17]) is a key factor in the development of products to maximize the customer value. The focus here is set on the integration of all relevant persons into the development process, with regard to a human-centred development, which is of great importance not only to the customer but also to the developer and the user. [10] In addition, ideas, concepts and prototypes are developed iteratively, validated with the customer (e.g. regarding their acceptance of the developed solutions) and further developed throughout the entire process in terms of the scope of functions and degree of maturity. [18] Already in the early stage of PGE a large number of decisions has to be made, which have a massive impact on the downstream development steps and later product properties. [14] In particular, it is important to make decisions based on the predicted adoption of future customers and users in order to strategically align the development of the product. The adoption process can be described as, the adoption by individuals, which in its entirety has already been introduced as diffusion. According to ROGERS, this process consists of five consecutive phases (Knowledge, Persuasion, Decision, Implementation and Confirmation). [8] Various influencing factors determine this mental process. [19] In addition to these factors, the interaction between adoption behaviour and a user's acceptance of the future product is highly relevant.

2.2 Acceptance regarding technical Connected Products

There is no uniform definition of acceptance; the meaning depends on the scientific point of view. Primarily, a distinction is made between the economic and sociological view-points, which leads to differing views on the phenomenon of acceptance. [20, 21] All definitions of the term “acceptance”, irrespective of the scientific field they originate from, contain synonyms such as "Adopt", "Acknowledge", "Affirm", "Agree", but also "Approve". [20–22] In general, however, acceptance does not represent a property, but rather the result of a reciprocal process of the subject-, the object- and the context of acceptance. [21] The subject of acceptance describes the initial point of acceptance (e.g. the user of the relevant product). The object of acceptance situates the acceptance. It describes the object which is accepted. The context of acceptance describes the environment in which the subject of acceptance interacts with the object of acceptance. In literature, so-called dimensions or levels of acceptance are discussed often. Overall, the dimensions of attitude, action and value (Fig. 2) are presented. The dimension of attitude is divided into two different components. [20] The affective component describes the emotional estimation, while the cognitive component contains an estimation based on the mind, which is usually grounded on a cost-benefit comparison. Overall, this dimension describes the decision on which parts are going to be carried over. If it is positive, it is called acceptance. It is important that it can only contain an intention to act, but not concrete action, because it is part of the dimension of action. In scientific jargon, this is also referred to as the conative dimension. It contains the behaviour of the user based on the dimension of attitude and can appear in different forms. Potential forms can be, for example, the purchase or the use of a technical object, but

IOP Conf. Series: Materials Science and Engineering **520** (2019) 012011 doi:10.1088/1757-899X/520/1/012011

also active resistance or absence of an action. [23] The normative dimension (value dimension) contains the attitude towards the object of acceptance based on existing norms and values, in some cases it is considered as part of the dimension of attitude.

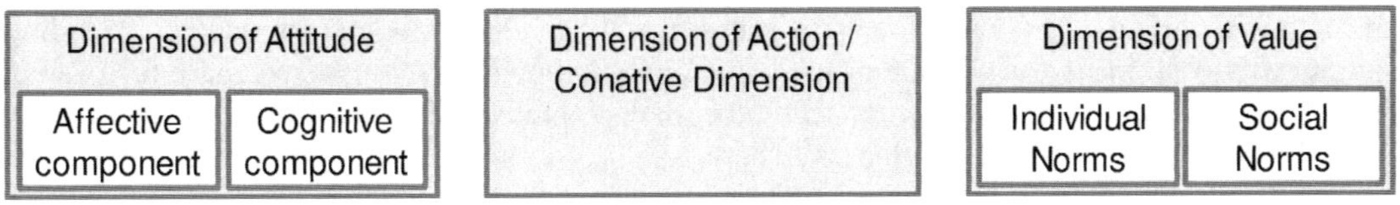

Figure 2. Overview of the different dimensions of acceptance

A distinction must be made between individual and social norms and values. [21] By introducing this dimension, it can be distinguished between the two forms of acceptance, adoption-acceptance and adaptation-acceptance. Both lead to the carry-over of the object, but are based on different circumstances. While in the case of adoption acceptance, all object functions fit into the existing value system and system of objectives and the object is transferred without external pressure, i.e. voluntarily. In the sense of adaptation acceptance, the object does not fit into the value system and system of objectives on which the normative evaluation is based and the object is therefore only transferred by external pressure. [20, 22]

The technical literature distinguishes between four different types of acceptance models. Input-Models are the simplest form of acceptance models. Within these, only the various factors, which influence the process of acceptance, are taken into account and, based on them, the formation of acceptance is presented. [22] Input-Output-Models do not only analyse the influencing factors, but also use them as input variables and additionally differentiate the result of the acceptance process. [22] An-other type are the so-called Feedback-Models, which have a process-related character. [22] They have the special feature that the influencing factors no longer appear as static dimensions, but change through feedback due to the result of the acceptance process. [20] The fourth category consists the Dynamic-Process-Models, which also do have the process-related character but as well are able to change. [22] One of the most common acceptance models is the TAM – Technology Acceptance Model (Fig. 3) by DAVIS [24]. It is designed as an Input-Model and is based on the theory of reasoned action. [22]

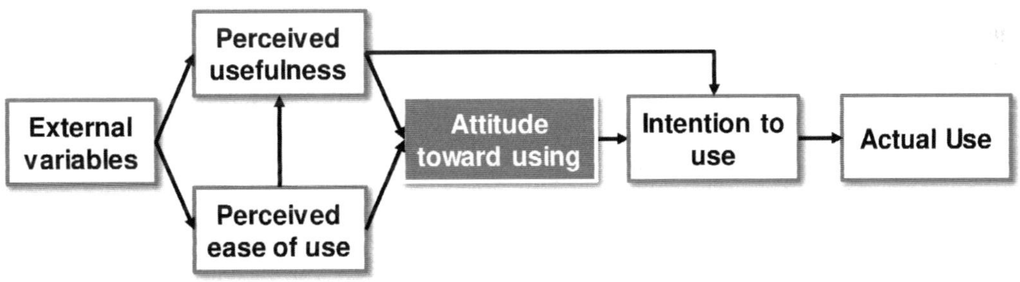

Figure 3. TAM-Technology Acceptance Modell in accordance with DAVIS [24]

The Attitude Toward Using is at the heart of this model. This in turn influences the Intention to Use, which has an influence on the Actual Use. An interesting point regarding the TAM is the separation between the intention to act and the actual action. On the influence side, importance is attached to the two variables Preceived Usefulness and Prceived Ease of Use. These two are in turn influenced by External Variables that are not further specified. Various studies have already demonstrated the usefulness of TAM, but have also shown that the External Variables need to be specified in more detail. [25, 26, 27] Later the TAM was expanded by DAVIS et al. [28] The focus was set on the two cluster of cognitively instrumental- and social variables to describe the input of the model more detailed. [28]

3. Research Approach

As already mentioned, the factor user-acceptance regarding the product to develop within the innovation process is a decisive factor for the sustainable competitive position of companies. If a technical system is not accepted by its users, no adoption will normally occur, irrespective of the other technical design or mode of operation, and success will therefore fail to materialize. In addition, decisions have to be made early in the process, although acceptance is still very vague, for example, whether the product development process should be implemented at all. These decisions in turn have a major influence on the subsequent process. There are already some useful acceptance-models, like TAM, but due to their domain-specific orientation [20], they are only conditionally suitable for use in mechatronic system development. Within the product development process, it is necessary, that the framework of the method is already given and the product developer only decides between various characteristics and receives easy interpretable results. This leads to the following, for this contribution relevant overriding research question: How does an acceptance model have to be designed that makes the construct of acceptance accessible to product developers in the early development stages? In addition, it is intended to identify potential acceptance barriers, but also opportunities with this model, which may accompany the product already in the early stage of PGE and enables the developer to derive recommendations for action with regard to the development strategy.

To answer this overriding question, the following sub-research-questions were derived, which at the same time represent the structure of the chapter of results in this work.

- What is the influence of user acceptance regarding connected products on adoption of products from the perspective of product development?
- Which modules are needed to model acceptance of individuals with regard to connected products?
- How to design an overall acceptance model for potential, future users to make connected products comparable within product engineering processes?
- Which recommendations for action can be derived based on the developed model in the ongoing product engineering process?

The procedure for the research work was therefore as follows:

1. Initially, based on literature research and studies relevant acceptance parameters were identified and then transferred to an initial acceptance model.
2. The overall acceptance model was specified by a networking analysis based on expert surveys and converted into a tool. (AMT - Acceptance Modelling Tool).
3. In a case study within a development process at Trumpf GmbH + Co. KG, the tool and the underlying model were continuously evaluated and refined.

4. Results

4.1. Influence of Acceptance on Adoption from the Product Development Perspective

In literature, there is no consensus whether acceptance is part of the dimension of action or whether it is attributed to the adoption process. [29] From a business management perspective, the acceptance initially only includes the readiness for the carry-over, and therefore is only expressed at the dimesion of attitude. The forms differ between acceptance and resistance. The decision of adoption is based on this and located within the dimension of action. It identifies itself by adoption or rejection. It is important to note that, although the willingness to carry-over, forms a base for the adoption decision, however acceptance does not guarantee adoption. [30] In order to make the concept of acceptance accessible to product developers, an approach based on tried and tested models is required. Within this article, the KANO-Model (Fig. 4) proposed in 1984, is used for this purpose. [31] Acceptance is given, when all basic acceptance-relevant needs are fulfilled from the user's point of view. User acceptance is strongly restricted, if the product does not meet one of these requirements.

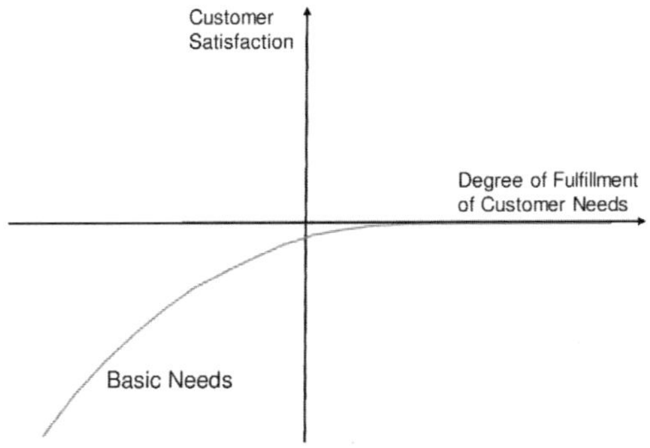

Figure 4. Simplified Model in accordance with KANO [31] – Basic needs represent the influence of Acceptance on Adoption

However, if the acceptance-relevant basic needs are fully met, customer satisfaction will not turn at all into enthusiasm, but only in a neutral attitude [31]. The requirements which are relevant for the respective acceptance, can be divided into the modules of acceptance, which are presented in the following chapter.

4.2. Modules of Acceptance

The construction of acceptance can be based on three modules, the object-, the subject- and the context of acceptance. (see 2.2) Acceptance is always object-related and therefore always focuses on an object as a reference point. Statements concerning acceptance always go hand in hand with the question: Acceptance of what? In principle, the object can be very diverse and may include for example technical products or political opinions. This diversity clearly shows that an acceptance analysis only makes sense with regard to a precisely defined object of acceptance. In the course of this publication, the Object of Acceptance is a connected product to integrate the manual workstation into an industry 4.0 strategy. The subject of acceptance forms the initial point of acceptance. It becomes the focal point by the question: Acceptance of whom? In this context, different characteristics can be distinguished as well. Thus, the subject can include individuals as well as social groups. Within this contribution the relevant subject of acceptance are the user of the connected product. The third module, the context of acceptance, describes the environment in which the subject gets in touch with the object. In principle, it includes all properties that are important for the acceptance process, but cannot be assigned to either the subject or the object. Acceptance is not a synonym for quality, but is rather described and defined by the interaction of the three previously introduced modules. [21] However, these three modules are too coarse-granular to make the construct of acceptance accessible to product developers. Therefore, it is important to describe them with so-called influencing factors. These were developed as part of this research on the basis of literature studies and expert workshops.

4.3. AMT – Acceptance Modelling Tool

The basis of the AMT – Acceptance Modelling Tool (Fig. 5) developed in the context of this contribution, which is strongly oriented towards the TAM - Technology Acceptance Model (See Fig. 3), was derived from the structure of an input-output-model. Heart of the model are the previously identified modules of acceptance and the alternating field they span. All information, necessary for the result of the acceptance process are defined via the input, which is formed by the sum of the influencing factors. The output of the model represents the result of the analysed acceptance process. It comprises the two parameter benefit, which the acceptance subject experiences through the use of the developed product and the change in its external impact within the social group (effect). The ratio of these two indicators reflects the result of the acceptance process.

DAVIS [24] sees the greatest influence on the Attitude Towards Using in the perceived usefulness (benefit). In addition, he attaches enormous importance to the social factors represented by the change in external impact (effect). [28] Due to that, those two quantities describe the output of the AMT.

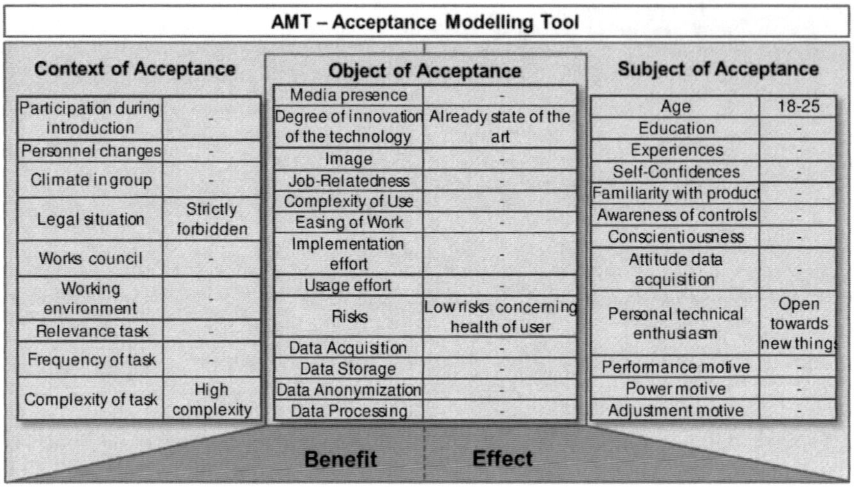

Figure 5. Simplified Overview of AMT, Modules of acceptance and chosen influencing factors, some with exemplary predefined characteristics

For a holistic description of the acceptance modules, various influencing factor were identified on the basis of a literature study [i.a. 25, 26, 27] and an expert workshop. It is important to note, that there are both, generic influencing factors, which are independent of the development context under consideration, and those, which are development context-specific. The tool, which supports the method described in this publication, contains all generic and development situation specific influencing factors and also offers predefined characteristics that the method-user can choose from. To describe the subject of acceptance, three different personae were developed, whose personal characteristics embody extremes (openness, scepticism and indifference) in the form of different user groups to represent potential future users of the developed product in the model. By selecting a persona, the corresponding proficiency level of the various persona-dependent influencing factors is automatically specified. The overall goal of the work behind this publication was to make the complex construct of acceptance quickly and easily accessible to the product developer. Therefore, the output of the model was quantified. The logic described in Section 4.1 has been stored for this purpose. As already mentioned, the user of the tool selects the characteristic of the corresponding influencing factors. Based on this, an assigned value is calculated on the basis of a stored value function. The function is based on the presentation of the basic requirements in the Kano-Model. Since, depending on personal characteristic of the subject of acceptance, different influencing factors also have different influences on the output (benefit and effect) of the model. This is why various persona-specific weightings for influencing factors are stored in the tool. Those weightings are based in an additional survey. Thus, when using the model, quantitative results of the acceptance process are obtained.

The result can be characterized under different aspects with regard to the aforementioned types of acceptance (adoption-acceptance and adaption-acceptance). While adoption-acceptance describes the voluntary adoption of an object, the adoption within the adaption-acceptance only happens because of an external pressure. If the subjectified, i.e. the persona-specific influence of an influencing factor on the output is selected and is compared with the subjectified reference, which is calculated from the persona-specific extremum of the influence-factor, an evaluation in the sense of adoption-acceptance is obtained. In this way, persona-unspecific opportunities and risks can be identified and based on this, object-focused development-priorities can be derived. However, if the subjectified influence is put in relation to the objectified reference, which is calculated from the absolute extreme of the influence over all personae, then it is an evaluation in the sense of the adaption-acceptance. This allows to compare different personae with regard to their acceptance toward the analyzed object, thus subject-specific recommendations for action can be derived.

4.4. Derivation of Recommendations

Based on the previously evaluated results, recommendations for action have been derived which support the product developer early in the process. In this context, four different fields of action were derived. The first three include the modules described above: object-, subject- and context of acceptance. As a fourth field, general strategies which basically lead to an increase of acceptance have been derived from a survey. In this context it is worth mentioning for example, that an adapted implementation process can promote transparency and thus generate trust. Official quality seals can also build up trust in the object of acceptance, i.e. in the networked system, and in its introduction. Another strategy involves creating additional incentives, for example by implementing a gamification approach. However, if the previously identified development priorities are to be met in a targeted manner, the other three fields of action are relevant. Development context-specific recommendations for action can be derived, if concrete opportunities or risks are identified based on the identified key indicators. For example, if the evaluation of adaptation-acceptance shows, that a persona is unsuitable for the corresponding development context, a change of persona can be forced. This corresponds to a change of the sales market or the place of use. A development potential concerning the context of acceptance is for example the integration of further functions and thus extending the application context. If the object of acceptance - i.e. the actual technical realization - serves as the focal point of consideration, a change in the relevant technology can be aimed at. If, in the course of this consideration, it is analysed that data collection in particular represents the risk, it is recommended that data will be made anonymous or at least be pseudonymised. In the application of the model, further, more concrete measures can be derived based on the opportunities and risks derived from the model, and thus user acceptance can be sustainably increased. It is crucial that the analysis can be carried out at an early stage of PGE. The potential weaknesses and opportunities, which are identified during the process, can thus be addressed as a potential basis for decision-making.

5. Evaluation of AMT in an Innovation Process

The acceptance model was developed as part of the project AIL – Agile Innovation Lab. [18] The focus of the development process, which was held in cooperation with Trumpf GmbH + Co. KG, was set on 'smart manual workstations in sheet metal processing'. Within the innovation process, a system was developed that analyses the work sequence of the operators of manual workstation and thus, among other things, carries out quality control and the necessary documentation. The application in the project enabled the evaluation of the AMT – Acceptance Modelling Tool. The AIL project is structured in time into the phases – Analyze, Identifying Potentials, Conception, Specification, Realization and Release – according to ASD – Agile System Design. [18] The tool was used during the phase of Identifying Potentials for the first time. In this context, the task was to describe the subject- and the context of acceptance initially in a holistic way. In addition, the different product profiles developed during this phase could be evaluated comparably with regard to acceptance by describing the object of acceptance within the AMT. Through the consistent usage of the model in this phase, the profiles and the selection process of these could be optimized with regard to acceptance. The generated insights were used to optimize the model. In the sub-sequent phases, the revised model was further used. In addition, in order to analyse the implemented changes, all previously developed product profiles were evaluated with the new development generation of the acceptance model to validate, whether the changes achieve the desired effect and, in addition the forecasted predictions of acceptance are compliant with the experts' estimates. This iterative approach, which extended over the entire innovation process, enabled the optimisation of the technical solutions regarding acceptance on the one hand and enabled the consequent evaluation of the acceptance model on the other hand.

6. Findings

Within the framework of research on which this publication is based, it was identified, that the result of the interplay between object-, subject- and context of acceptance reflects the result of the acceptance process. In order to model user-acceptance, it is necessary to analyze this interplay precisely, but as well to quantify the previously gained knowledge regarding the influence of acceptance on adoption in terms of product development. In this context, the approach of the KANO-Model was used. According to this approach user's acceptance of a product is given if all acceptance-relevant basic requirements are met from the point of view of the considered respective user segment. While modelling the acceptance, it was possible to gain insights that the benefit and the effect that the acceptance subject experiences through the use of the acceptance object in a defined acceptance context can be regarded as the acceptance result and thus make the complex construct of acceptance accessible to the product developer. On the basis of various previously identified influencing factors and their characteristics, which are selected according to the development context, key figures could be derived. These support the identification of development priorities. According to this, various recommendations for action were obtained, which can be divided into four different fields. A first area includes common recommendations, which can be applied generally to increase acceptance. The other three areas include recommendations that can be assigned to the three modules of acceptance and eliminate or promote corresponding risks and opportunities. In addition, various alternative solutions can be characterized and compared with regard to the expected user acceptance. Thus, uncertainties that exist in the early stage of the PGE can be resolved and necessary decisions can be made on a sustainable basis.

7. Future Works

To gain a deeper understanding of acceptance, the developed model will be used in further ASD Live-Labs. In the course of this process, further insights, which are necessary for a deeper interpretation of the results, can be acquired with regard to the key indicators. In particular, the acceptance of new technologies is highly dynamic. It will be crucial to continuously analyse, adapt and display the future user and his acceptance. The integration of a monitoring system of acceptance factors and a concept for "intentional forgetting" of assumptions that are no longer current are therefore seen as the next steps during the innovation process.

Acknowledgments

We would like to thank Trumpf GmbH + Co. KG and especially the team around Dr. Jens Ottnad for the collegial and open cooperation.

References

[1] Radziwon A, Bilberg A, Bogers M and Madsen E S 2014 The Smart Factory: Exploring Adaptive and Flexible Manufacturing Solutions, *Procedia Engineering*, **69**, 1184–1190
[2] Bauernhansl T, Hompel M and Vogel-Heuser B 2014 Industrie 4.0 in Produktion, Automatisierung und Logistik, Springer Fachmedien Wiesbaden
[3] Wolter M, Mönnig A, Hummel M, Schneemann C, Weber E, Zika G, Helmrich R, Maier T and Neuber-Pohl C 2015 Industrie 4.0 und die Folgen für Arbeitsmarkt und Wirtschaft: Szenariorechnungen im Rahmen der BIBB-IAB-Qualifikations- und Berufsfeldprojektionen, *IAB-Forschungsbericht*
[4] Ganschar O, Gerlach S, Hämmerle M, Krause T and Schlund S 2013 Produktionsarbeit der Zukunft - Industrie 4.0, Spath D, editor. Stuttgart: Fraunhofer Verlag
[5] Wirtschaftswoche – Report about Hannovermesse [Internet], Available from https://www.wiwo.de/unternehmen/mittelstand/hannovermesse/industrie-4-0-so-sieht-der-arbeitsplatz-der-zukunft-aus/12583554.html [Accessed 2018-10-20]
[6] Ittermann P, Niehaus J and Hirsch-Kreinsen H 2015 Arbeiten in der Industrie 4.0: Trendbestimmungen und arbeitspolitische Handlungsfelder, *Studie der Hans-Böckler-Stiftung*
[7] BITKOM 2011 Soziale Netzwerke: Eine repräsentative Untersuchung zur Nutzung sozialer Netzwerke im Internet. 2nd ed.
[8] Schumpeter J 1939 Business Cycles: A Theoretical, Historical and Statistical Analysis of the Capitalist Process. New York: McGraw-Hill

[9] Rogers E 2010 Diffusion Of Innovations, 4Th Edition

[10] Albers A, Heimicke J, Walter B, Basedov G N, Reiß N, Heitger N, Ott S and Bursac N 2018 Product Profilers: Modelling customer benefits as a foundation to bring inventions to innovations, *Procedia CIRP*, **70**, 253-258

[11] Cooper R and Kleinschmidt E 1987 Success factors in product innovation. Industrial Marketing Management, **16**, 215–223

[12] Definition » Diffusion « | Gabler Wirtschaftslexikon [Internet]. Available from: http://wirtschaftslexikon.gabler.de/Definition/diffusion.html [Accessed 2018-10-25]

[13] Albers A, Bursac N and Rapp S 2017 PGE – Produktgenerationsentwicklung am Beispiel des Zweimassenschwungrads, *Forschung im Ingenieurwesen*, **81**, 13–31

[14] Albers A, Rapp S, Birk C and Bursac N 2017 Die Frühe Phase der PGE – Produktgenerations-entwicklung, *Proceedings of Stuttgarter Symposium für Produktentwicklung*

[15] Albers A, Reiß N, Bursac N and Richter T 2016 iPeM – Integrated Product Engineering Model in Context of Product Generation Engineering, *Procedia CIRP*, **50**, 100–105

[16] Albers A, Reiß N, Bursac N, Walter B and Gladysz B 2015 InnoFox – Situationsspezifische Methodenempfehlung im Produktentstehungsprozess, *Proceedings of Stuttgarter Symposium für Produktentwicklung*

[17] Prahalad C K, Ramaswamy V 2002 The Co-Creation Connection. Strategy and Business, 50–60

[18] Albers A, Bursac N, Heimicke J, Walter B, Reiß N 2017 20 years of co-creation using case based learning: An integrated approach for teaching innovation and research in Product Generation Engineering, *Auer, Guralnick et al. (Ed.) 2017 – Proceedings of the 20th ICL*, 636-647

[19] Albers S 2001 Marktdurchsetzung von Innovationen, *Technologie- und Innovationsmanagement: Leistungsbilanz des Kieler Graduiertenkollegs*, 1st ed., 79–116.

[20] Kollmann T 1998 Akzeptanz innovativer Nutzungsgüter und -systeme: Konsequenzen für die Einführung von Telekommunikations- und Multimediasystemen

[21] Lucke D 1995 Akzeptanz Legitimität in der „Abstimmungsgesellschaft“

[22] Schäfer M and Keppler D 2013 Modelle der technikorientierten Akzeptanzforschung: Überblick und Reflexion am Beispiel eines Forschungsprojekts zur Implementierung innovativer technischer Energieeffizienz-Maßnahmen

[23] Huijts N, Molin E and Steg L 2012 Psychological factors influencing sustainable energy technology acceptance: A review-based comprehensive framework, *Renewable and Sustainable Energy Reviews,* **16**, 525–531

[24] Davis F D 1989 Perceived usefulness, perceived ease of use, and user acceptance of information technologies, *MIS Quarterly*, **13 (3)**, 319–340

[25] Legris P, Ingham J, Collerette P 2003 Why do people use information technology? A critical review of the technology acceptance model, *Information & Management*, **40**

[26] Chang C C, Yan C F, Tseng J 2012 Perceived convenience in an extended technology acceptance model: Mobile technology and English learning for college students, Aus. J. of Edu. Tech.

[27] Lee Y, Kozar K A and Larsen KRT 2003 The Technology acceptance model: past, present, and future, Com. of the Ass. for Info. Sys., **12**, 752-780

[28] Venkatesh V and Davis F 2000 A Theoretical Extension of the Technology Accptance Model: Four Longitudinal Field Studies, Man. Sc., **46**

[29] Reichardt T 2008 Bedürfnisorientierte Marktstrukturanalyse für technische Innovationen: Eine empirische Untersuchung am Beispiel Mobile Commerce

[30] Binsack M 2003 Akzeptanz neuer Produkte. Wiesbaden: Deutscher Universitätsverlag

[31] Kano N, Seraku N and Takahashi F 1984 Attractive Quality and Must-be Quality, J. of the Jap. Soc. for Q. Ctrl, **14**, 147–15

The Construction of Visual Cognitive Decision-making Platform with User Insight and The Fashion Trend in the Era of Big Data

Yujing Tian[1], Ziran Zhang[2,*], Ying Yu[1], Xiaogang Liu[1] and Huiyin Huang[3]

[1] Donghua University, Fashion and Design College, 200051 Shanghai, P.R. China
[2] Shanghai University of Engineering Science, School of Art and Design, 201620 Shanghai, P.R. China
[3] Shanghai Tianhao Information Technology Co., Ltd., 200120 Shanghai, P.R. China

* zhangziran@sues.edu.cn

Abstract. With the rapid development of the Internet, big data have brought better tools and perspectives for the research and development of precise fashion products. In this paper, based on the fashion design industry, interactive operation principle and consumer demand, the relevant data are integrated and the innovative technology architecture is utilized, a decision-making system for data visualization is constructed, and the structural problems among fashion trend prediction, market sales and consumer behavior are accurately described so as to realize accurate fashion trend prediction and user behavior insight, help to predict the precise fashion trend of the brand and complete the rapid recognition of consumers and the adjustment of design strategy.

1. Introduction

Driven by the background of innovation and transformation in China's fashion industry, the trend prediction of most fashion enterprises still tends to the perceptual judgment of decision makers, and this makes the decision-making work more random in objectivity, scientificity and accuracy[1]. Without accurate guidance and scientific product decision-making and early warning mechanism, product development severely separates the intrinsic relationship among popular trends, product strategies and consumer demands, thereby resulting in blind and disorderly product development and market release, and a large amount of inventory. However, essentially, as a data-driven business form centered on consumer experience, new consumption integrates relevant data of prediction, market and user around consumer demand, innovates technological framework, constructs a large data closed-loop and visualization platform of fashion trend from trend prediction to product, market and consumer behavior, and helps enterprises to complete the judgment of fashion trends, consumers' quick recognition and design strategy adjustment, flexible product lines, and effectively reduce the inventory of fashion industry.

2. Research on Visualization System of Fashion Trend and Consumer Behaviour Insight

The core of fashion trend and consumer behavior insight decision-making platform based on data visualization is to collect and extract all kinds of real-time sales, consumer and fashion trend data reflected in the design and development process, and use mature mathematical models to accurately calculate and authentically describe the status of trend prediction and consumer behavior, and display the results to brand/design decision makers through the visualization software platform system. This is a complex system project. In other fields, visualization is more familiar, but it has not been widely used in the fashion industry, involving computer information technology, data visualization software platform technology, and fashion prediction technology[2]. The specific situation is described below.

2.1. Construction of Basic Data System

The premise of data visualization is the collection, analysis and application of fashion trends, user portraits and marketing data, which can be divided into three categories:

The first one is the sales data of the macro market, including sales volume, colour, channel, style, promotion, region and price. The next one is the data described by consumers, including basic characteristics, social contact, behaviour, interest, psychology and income. The last one is the data oriented to fashion trends[3-4], including events, materials, colours, styles and popular elements. Through the collection, collation, induction, analysis and prediction of massive data, the causality behind various kinds of data is excavated from complex data, and various mature modeling techniques are applied to promote these basic data from quantitative to qualitative changes, so as to realize the deep application of massive basic data and timely presentation of data and support brand/design decision makers to make scientific decisions[5]. As shown in Figure 1.

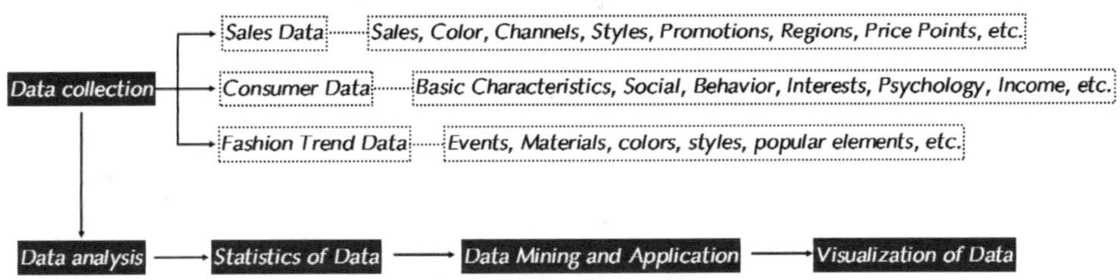

Figure 1. Application of Basic Data.

2.2. The Connotation and Essence of Visualization Research

Instead of simply displaying the big data system of fashion trend and sales system operation to brand/design management decision makers through data visualization software platform, data visualization is to reassemble, analyze and deep mine the basic data system through complex data calculation, analysis and fitting, so as to truly reflect the value of data. The fashion trend and the connotation and essence of consumer behaviour insight data visualization can be divided into the following levels:

- The first is the presentation of key market sales data, which mainly reflects the overall situation of market sales, the trend forecast and sales situation of product attributes (style and colour), and the distribution and layout of consumers. Through the presentation of diverse data, it provides an intuitive decision-making basis for brand/design management decision makers.
- The second is the deep correlation of all kinds of data. The data of market, product, and trend and consumer behaviour are deeply integrated and correlated, and the integration of multi-source data is used to dig out the deep relationship among all kinds of data, so as to form a closed loop of fashion trend big data and find the potential rules.
- The third is the application of all kinds of key data and rules. All kinds of key data and rules are summarized to give alternative decision-making ideas for brand/design management decision makers and provide support for the final management decision.
- The fourth is to realize the interactive experience between brand/design management decision makers, data presenters, fashion trends and consumer behaviour insight visualization platform so as to form a smooth and real-time information sharing channel among the three, break through the barriers from the data level to realize the data closed loop of the whole industrial

chain, and establish an efficient brand/design decision cycle mechanism. The business model is shown in Figure 2.

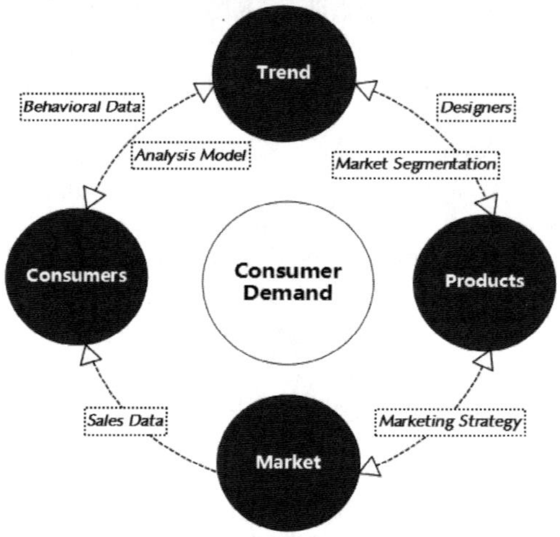

Figure 2. Business Model of the System.

2.3. Service Mode of Visualization Platform

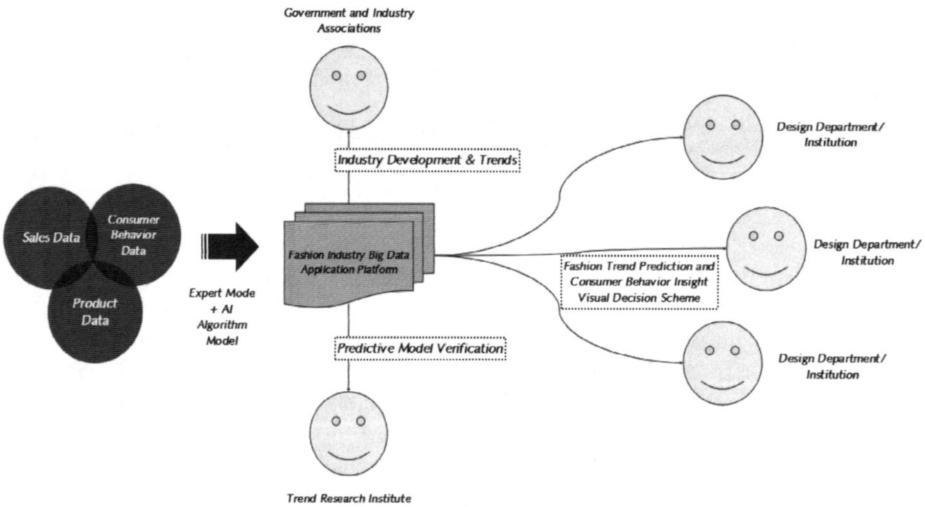

Figure 3. Service Mode of the System.

The main users of the system include brand trend research department, brand product design decision-making department, government and industry associations. Through user interviews, they hope that the visual decision-making platform can help them show their prediction results to high-level decision makers in real time and in an intuitive way. Their needs mainly focus on three aspects: Firstly, for trend research departments or institutions, they hope to improve the accuracy of trend research and influence other related departments by assisting them to provide data verification of prediction results through prediction models[6], especially the design strategy of the design and development department. Secondly, for enterprises or brands, they hope to monitor the trend prediction and sales results in a real-time and whole-process manner, guide consumers to complete the rapid cognition and behavioural insight, so as to adjust the error between trend prediction and product direction, define the design strategy, and improve product sales. Thirdly, for the government and industry associations, they hope that the big data application platform of the fashion industry can provide direction and trend guidance for the industry development.

According to the above needs analysis, the visualization platform needs to integrate sales data, product data and consumer behaviour data, and construct a large data application platform for fashion industry through expert model and AI algorithm model, accurately describe and solve the structural problems between brand product design and fashion trend prediction, design positioning and consumer demand, purchase behaviour and fashion trend research. The service model of the system is shown in Figure 3.

3. Design of Decision-making Platform Based on Fashion Trend and Consumer Behaviour Insight

Based on the real data provided by a fashion luggage brand, the distributed processing platform is used to pre-process and store the accessed data. According to the function orientation of the decision-making platform, the key decision-making index data based on the needs of the fashion industry are listed, and the visual cognitive decision-making platform for fashion trend and user insight is designed.

3.1. Function Setting and Key Index Extraction of the System

3.1.1. Real-time Sales Situation Description Function Real-time sales data of brand/product are described and displayed dynamically, and the overall operational situation is accurately described. After screening, it mainly includes cumulative sales and year-on-year situation, cumulative sales and year-on-year situation, annual index, annual index and completion status, average unit price, proportion of price and sales, monthly sales volume, best-selling colour, best-selling style and proportion of cities and sales, etc., as shown in Table 1.

Table 1. Key indicators related to sales.

Main indicators	Definition of Indicators	Specific description and examples
Cumulative sales and year-on-year situation	Accumulated Sales and Year-on-Year Analysis of Brands	XX (Unit: 10,000 RMB), up/down XX% year on year
Accumulated sales and year-on-year situation	Accumulated Sales of Brands and Analysis of Year-on-Year Situation	XX pcs, up/down XX% year on year
Annual Indicators and completion status	Brand Annual Indicators and Completion Analysis	XX (Unit: 10,000 RMB), completed XX%
Average unit price	Average unit price of products sold by brands	XX RMB, up/down XX% year on year
Proportion of price and sales	Proportion and distribution of various sales price ranges	Proportion below 350, 350-450, 450-550, 550-650, 650-750, 750-850, 850-950, and above 950
Monthly sales volume	Monthly sales in recent two years	Monthly sales from January to December 2017 and 2018
Best-selling colour	Top ten colours of sold products	Black, brown, red, white, orange, camel and pink-orange colours and their proportion
Best-selling style	Main styles and proportions of sold products	Styles and proportions of shoulder bag, S-M bag, handbag, messenger handbag, messenger bag, chest bag, backpack, dinner bag, handheld shoulder bag, etc.
Proportion of cities and sales	The cities where the brands are mainly sold and the proportion of sales volume	Guangzhou, Shanghai, Beijing, Chengdu, Shenzhen and their sales proportion

3.1.2. Fashion trend prediction and sales status comparative analysis function By dynamically focusing on sales data such as product style, color and the proportion of city sales, comparing the difference between trend prediction content and actual sales, real-time fashion prediction bias warning function, and guiding the discovery and prediction of discrepancies, the relevant data behind a SKU and a certain color under the problem point are tracked, including the basic sales situation (cumulative sales and year-on-year situation, cumulative sales volume and year-on-year situation), the comparative analysis of trend forecast and current situation (age comparison, income comparison, monthly sales volume comparison, sales proportion and distribution in cities), and the sales data links of similar products. As shown in Table 2, the prediction results of a popular color are compared with the current sales situation by labeling the following indicator data values of different colors.

Table 2. Main indexes for comparing the current situation with the prediction of a popular color.

First-level index	Second-level index	Definition of Indicators	Specific description and examples
A popular color (Take yellow for example)	Yellow bags Best-selling styles TOP 1 to TOP 5	Cumulative sales and year-on-year situation	The predicted situation is marked in blue The actual situation is marked in yellow Contrastive analysis
		Cumulative sales volume and year-on-year situation	
		Age comparison	
		Income comparison	
		Monthly sales comparison	
		Sales proportion and distribution in cities	

3.1.3. Consumer Behavior and Multidimensional Dynamic Portrait Function By tracking the sales data of SKU and each series collection in real time, and updating the consumers behind the sales data in real time, the dynamic and visual user portrait can be formed finally. The content of the portrait includes the basic information of consumers (including gender, age, income situation, new and old customers) and consumer behavior habits (including discount sensitivity, browsing preferences, purchase preferences, search adjectives, etc.), as shown in Table 3.

Table 3. Key Indicators of User Portrait.

First-level index	Second-level index	Definition of Indicators	Specific description and examples
Basic information of consumers	User gender	Gender composition of users	Male, female, unknown
	Age	Age and composition of users	18-25, 26-30, 31-35, 36-40, 41-50, above 50, unknown
	Income situation	Income and proportion of users	3k-5k, 5k-8k, 8k-10k, 10k-20k, above 20K
	New and old customers	Number and proportion of new and old customers	New and old customers
Consumer behavior habits	Discount sensitivity	Consumers' sensitivity to discounts can assist in formulating appropriate pricing strategies.	High, medium, low and unknown
	Browsing preferences	Consumers' browsing	Luggage leather/Hot selling

	preferences help to configure promotion or cross-border cooperation in relevant preference categories	women's bag/Men's bag, Women's wear/ boutique, Women's shoes, Women's underwear/Men's underwear/Home wear, Makeup and skin care/Body care/Essential oil, etc.
Purchase preferences	The purchasing preference of consumers facilitates to adjust the width and depth of product line	Women's bags, low-heel shoes, dresses, T-shirts, trousers, mobile phone accessories, wallets, sweaters, China Mobile recharge cards, shirts, etc.
Search adjectives	The adjectives that consumers search for help match the relevant goods under the relevant adjectives	Bag-female, canvas bags, ladies' bags, Little bag- female, ladies' bags 2019, handbags, messenger bags, new shoulder bags 2019, ladies' messenger bags, etc.

3.2. Technical Architecture of the System

Traditional visual data analysis is based on A, B, C, D... The system collects data from the bottom to the top, stores data through data storage, distributes data through Hadoop, and data mining with Mahout through customized programs, and finally presents a visual interface.

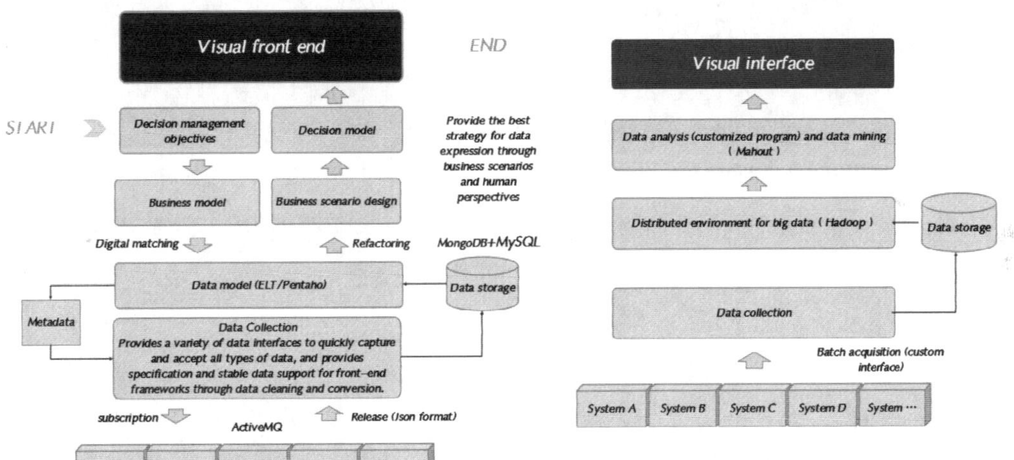

Figure 4. Comparison of Innovative Technical Architecture and Traditional Visual Data Analysis System.

The technical framework of this paper is that the decision-making system based on data visualization should accurately describe and solve the structural problems between brand product design and fashion trend prediction, design positioning and consumer demand, purchase behavior and fashion trend research according to the interactive operation principle, using the standardized model base of each system function.

Based on the target analysis of decision management, after deeply laying out the business model, the data models from A, B, C, D are completed by ELT/Pentaho, etc. The metadata of different data interfaces of the system is collected and cleaned by software, which can be converted into front-end framework to provide normative and stable data support. The distributed processing platform is used

IOP Conf. Series: Materials Science and Engineering **520** (2019) 012012 doi:10.1088/1757-899X/520/1/012012

to pre-process and store the accessed data, and the key decision-making index data based on the needs of the fashion industry are obtained. The business scenario design and decision-making model are used to import the front-end of shaping visualization, as shown in Figure 4.

3.3. Visual presentation of the system

After all the data presentation, data association and data application of the decision platform are completed, the last step is to realize the visualization of the platform. Based on the data of fashion luggage brand, the visualization platform of fashion trend and consumer behavior is formed, which is shown in Figure 5-7. The real-time sales situation, fashion trend prediction, comparative analysis of sales status, consumer behavior and multi-dimensional dynamic portrait are realized.

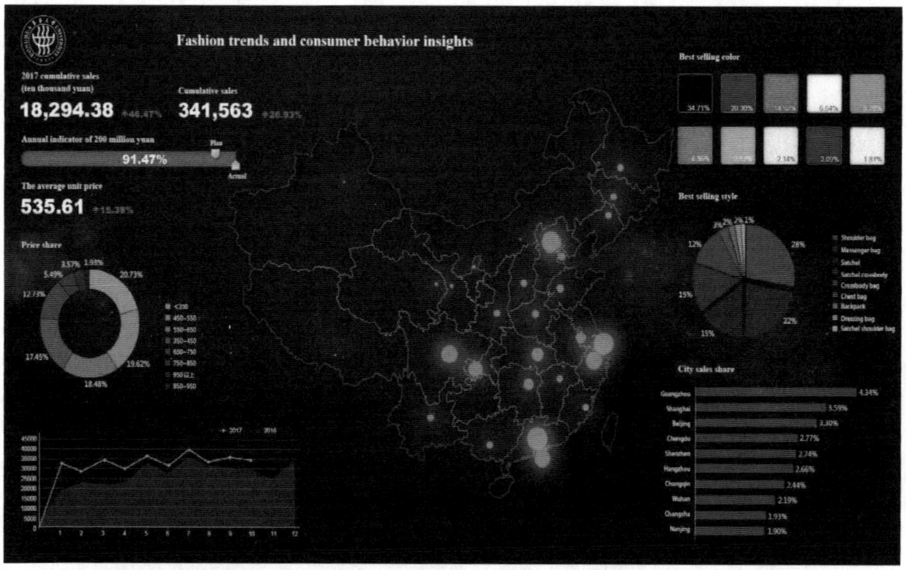

Figure 5. Visual presentation illustrations of real-time sales situation

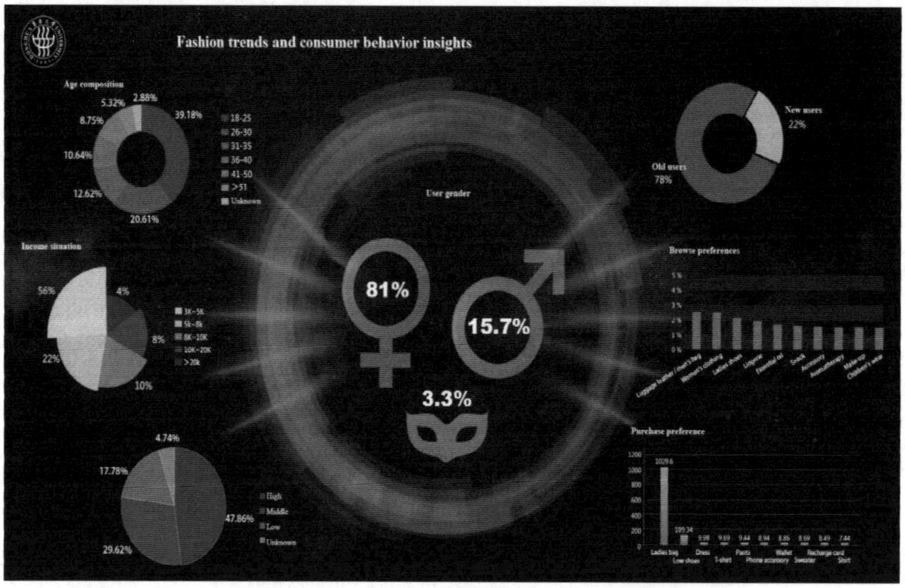

Figure 6. Visual presentation illustrations of consumer behavior insight

IOP Conf. Series: Materials Science and Engineering **520** (2019) 012012 doi:10.1088/1757-899X/520/1/012012

Figure 7. Visual presentation illustrations of fashion trends comparative analysis of sales status

4. Conclusions

The construction of user behavior insight and visual decision-making platform for fashion trend calibration based on big data is a complex system which combines many new technologies. By combining fashion trend prediction data, user's basic data, real-time sales data and decision-making management objectives, the data closed-loop is realized through opening barriers from the data level, and the massive data are effectively utilized and viewed. User portraits, consumer behavior and fashion trends are presented sensitively in order to provide great convenience.

Acknowledgments

Project Support by the Fundamental Research Funds for the Central Universities, Shanghai Style Fashion Design &Value Creation Collaborative Innovation Center & the Fundamental Research Funds for the Central Universities & Shanghai Summit Discipline in Design (DB18212)

References

[1] Hu L 2015 *Chin. Text. Lead.* 78-80
[2] Wu J and Zheng S 2015 *J. Comput. Appl.* **35** 146-50
[3] Alhalah Z, Stiefelhagen R and Grauman K 2017 *Int. Conf. on Computer Vision* (Venice: IEEE Computer Society)
[4] He R, C. Lin C and Mcauley J 2016 *Int. Conf. Companion on World Wide Web* (Montreal)
[5] Corbiere C, Ben-Younes H, Rame A and Ollion C 2017 *Int. Conf. on Computer Vision Workshop* (Venice: IEEE Computer Society)
[6] Wang W and Ma C 2015 *Tianjin Text. Sci. Technol.* **3** 51-3

IOP Conf. Series: Materials Science and Engineering **520** (2019) 012013 doi:10.1088/1757-899X/520/1/012013

Mechanical Modeling the Processes of Additive Manufacturing Bearing Pillars of Viscoelastic Aging Materials by Thickening over the Lateral Surface under Axial Loading

K E Kazakov[1,*] and D A Parshin[1,2]

[1] Ishlinsky Institute for Problems in Mechanics of the Russian Academy of Sciences, Vernadsky Ave. 101 Bldg 1, Moscow, 119526, Russia
[2] Bauman Moscow State Technical University, 2nd Baumanskaya Str. 5/1, Moscow, 105005, Russia

*E-mail: kazakov-ke@yandex.ru

Abstract. A nonclassical mechanical model for the process of additive manufacturing a bearing circular conical pillar by means of its layer-by-layer thickening over the lateral surface using a viscoelastic aging material under simultaneous action of an axial variable central force is developed. The exact in the sense of Saint-Venant principle analytical solution of the stated corresponding quasi-static boundary value problem is constructed. The obtained solution allows to retrace the evolution of the stress-strain state of the manufactured pillar after the start and on completing the piecewise continuous additive process under consideration.

1. Introduction

Technological processes of additive manufacturing elements of structures are accompanied by an increase in the size and shape of the solids obtained in these processes due to the addition of new material layers to their surfaces. A distinctive feature of such kind of processes is the inherent fact that the considered solid does not exist in its final form before the beginning of its deformation process, as it is always considered in the classical mechanics of deformable solids: the solid is formed already in the course of this process. Therefore, in the mechanical analysis of additive processes, it is necessary to simultaneously take into account both external loads acting in the simulated technological process on the solid under consideration, including loads acting on the additional material being attached to it, and the geometric, kinematic and force characteristics of the process of gradual attachment of new material elements to the solid. Due to the absence in principle for the entire additive-manufactured solid any unstressed configuration which could act as a reference configuration for traditional strain measures, the mentioned account cannot be correctly carried out within the framework of classical solid mechanics, even if the traditional equations and conditions are formulated for a time variable spatial domain. Thus, the problems on modeling mechanical behaviour of additive-manufactured solids should have a number of specific features and constitute a special class of problems of solid mechanics. The state-of-the-art of the corresponding nonclassical field of mechanics and particular results achieved in the framework of Russian science school in mechanics of growth created by Professor Manzhirov can be examined, for example, on the basis of works [1–7]. The present study is devoted to a nonclassical quasi-static problem for an additively manufactured solid formulated with the satisfaction of static boundary conditions on a parts of the solid boundary surface in the integral sense. The main obstacle to construct the solution of such problems is the time expansion of the surface part with the integral conditions due to influx of the additional material to another surface part adjacent to the former one. This obstacle is overcome in the study.

2. Material relations for the viscoelastic aging material used

Many artificial and natural materials used in additive manufacturing processes exhibit pronounced rheological properties, in particular, the properties of deformation heredity (viscoelasticity) and aging (weakening over time the deformation properties of the material regardless of the mechanical loads acting on it). Additive processes in which solids are formed from such materials are quite difficult to simulate, since the time-evolving deformation reaction of such a solid to the impacts applied to it continuously interacts with its mechanical response to the process of adding new material elements, and during pauses in the growth process, as well as after the final termination of this process, the formed solid continues to change its stress-strain state. However, the study of this kind of processes is relevant from the point of view of a variety of engineering and physical applications.

We will consider the viscoelastic aging material described by the following linear isotropic relationship between the stress tensor $\mathbf{T}$ and the small strain tensor $\mathbf{E}$ at arbitrary point $\mathbf{r}$ of the considered solid at the time instant t ($\mathbf{1}$ is the unit tensor of the second rank) [8]:

$$\mathbf{T}(\mathbf{r},t) = G(t)\left[\mathbf{T}^{\circ}(\mathbf{r},t) + \int_{\tau_0}^{t} \mathbf{T}^{\circ}(\mathbf{r},\tau)R(t,\tau)\,\mathrm{d}\tau\right] \tag{1}$$

where we use the notation

$$\mathbf{T}^{\circ}(\mathbf{r},t) = 2\mathbf{E}(\mathbf{r},t) + 2\nu(1-2\nu)^{-1}\mathbf{1}\mathbf{1}\bullet\bullet\,\mathbf{E}(\mathbf{r},t)\,. \tag{2}$$

The kernel $R(t,\tau)$ is called relaxation kernel and is the resolvent of the creep kernel $K(t,\tau)$ which can be determined by the expression

$$K(t,\tau) = G(\tau)\,\partial\Delta(t,\tau)/\partial\tau\,. \tag{3}$$

The quantity $G(\tau)$ is the elastic shear modulus increasing with the material age $\tau > 0$ (we calculate the material age from the material origination instant) due to aging of the material; ν is the constant Poisson ratio for elastic and creep strain. The so called specific strain function for pure shear $\Delta(t,\tau)$ describes the actual total strain, elastic and creeping, to the time instant $t \geq \tau$ of a material specimen under the constant unit pure shear loading which is acting starting from the time instant $\tau > 0$:

$$\Delta(t,\tau) = G(\tau)^{-1} + \omega(t,\tau) \tag{4}$$

where $\omega(t,\tau)$ is the creep measure for pure shear, $\omega(\tau,\tau) \equiv 0$, see figure 1. It is to be noted that the quantity $\omega(+\infty,\tau)$ has meaning of creep resource of the material at its age $\tau > 0$. This material characteristic decreases with the age augmentation due to the aging process going on regardless of any loadings acting on the material. The creep kernel $K(t,\tau)$ can be also expressed through other characteristics of the material, describing its behavior in different elementary stress states. Given expression (3) uses a variant with the characteristics for the elementary state of pure shear.

As we can see, state equation (1) contains as a specific case the state equation of isotropic elastic material. The latter equation follows from (1) by taking $\omega(t,\tau) \equiv 0$, $G(\tau) \equiv \mathrm{const}$.

The time instant τ_0 in (1) is the instant of occurrence of stresses at the solid points. If state equation (1) is used to describe the mechanical behaviour of a growing solid which is formed additively by attaching new material layers to the current solid surface then the time τ_0 of stresses occurrence must depend on the variable radius-vector $\mathbf{r}$ of the solid points: $\tau_0 = \tau_0(\mathbf{r})$. As in the considered growing process the additional material is being loaded simultaneously with its attaching to the solid, the corresponding function $\tau_0(\mathbf{r})$ is to be determined in the following way. In the original part of the considered conical pillar this function should be identically equal to the time moment t_0 of loading of this part, and in the formed due to growth additional part of the solid under consideration this function should coincide with the distribution $\tau_*(\mathbf{r})$ of moments of attaching of particles $\mathbf{r}$ of the additional material to the solid.

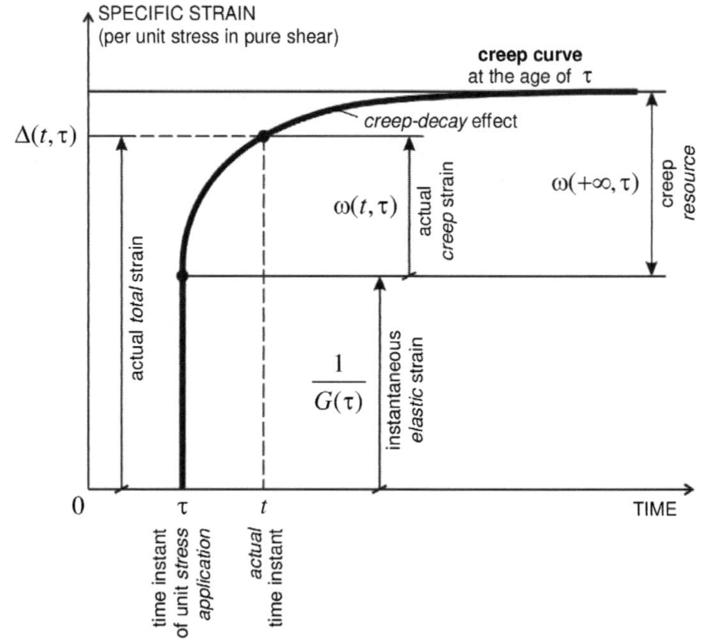

Figure 1. A conceptual creep curve of the used viscoelastic aging material.

Remark that constitutive equations similar to (1) are widely used to describe the mechanical behavior of various natural and artificial stones (in particular, concrete), polymers, soils, ice, wood. Typical experimental creep curves for such materials can be borrowed, e.g., from [9].

3. The mechanical problem under consideration

Let us consider a conical significantly high axisymmetric pillar made from the material subordinated to constitutive equation (1). At the moment $t = t_0 > 0$ an external loading is applied to the upper (bearing) end of the pillar. We assume that at every moment of time $t \geq t_0$ this loading is statically equivalent to a central axial force $P(t)$ which can arbitrarily vary with the time t. A compressing force $P(t)$ will be considered as positive.

At the instant $t = t_1 \geq t_0$, some time after the loading application, the process of gradual, layer-by-layer, axisymmetric thickening the considered pillar is starting. Thickening is implemented via piecewise-continuous adding the additional material to the pillar lateral surface. The added material is identic to the material of the originally existing pillar and is initially free of any stresses. Thickening occurs in such a way that at each time moment t the pillar maintains a conical shape. We denote the with time increasing radius of the bearing end of the pillar by $a(t)$, and we denote $a_0 = a(t_0) = a(t_1)$. The current angle $\alpha(t)$ of the pillar tapering can arbitrarily change due to pillar thickening. The model of piecewise-continuous thickening supposes that the process consists of $N \geq 1$ consecutive stages $t \in (t_{2k-1}, t_{2k})$, $k = \overline{1, N}$, of the pillar continuous accreting. These arbitrary long stages are separated by pauses $t \in [t_{2k}, t_{2k+1}]$, $k = \overline{1, N-1}$, of arbitrary duration when the influx of additional material to the pillar lateral surface does not take place. At the stages of continuous accreting an infinitely thin layer of additional material attaches to the lateral pillar surface each infinitely small period of time. In the whole process of piecewise-continuous thickening, $t \in (t_1, t_{2N})$, and after this process completion, $t \in [t_{2N}, +\infty)$, a time-varying central axial force $P(t)$ continues to act on the bearing end of the pillar. The modelled process is schematically illustrated in figure 2.

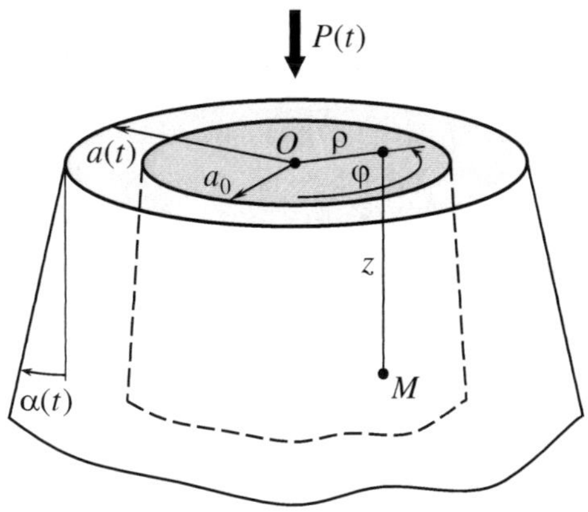

Figure 2. A scheme of the modelled process. The coordinate system used.

We investigate the evolution of the stress-strain state of the considered pillar under specified loading before the start, during and upon the completion of the described piecewise-continuous process of its thickening in the case of small strain. The deforming process is assumed quasi-static.

We place the origin O of coordinates into the geometric center of the bearing end of the pillar. We direct the coordinate axis Oz perpendicular to this end inside the pillar. Denote the polar radius and the polar angle of the polar cylindrical coordinate system with the reference axis Oz by ρ and φ respectively (see figure 2). Let the vector system $\{\mathbf{e}_z, \mathbf{e}_\rho(\varphi), \mathbf{e}_\varphi(\varphi)\}$ be the local orthonormal basis of the introduced cylindrical coordinate system (z, ρ, φ). It is clear that the radius-vector $\mathbf{r} = \overrightarrow{OM}$ of an arbitrary point M of the considered solid can be represented in the form $\mathbf{r} = \mathbf{e}_z\, z + \mathbf{e}_\rho(\varphi)\rho$.

One can describe changing the geometry of the considered circular conical pillar due to its thickening over the lateral surface by defining the shape function $\rho = L(z,t)$ for the current lateral surface of the cone: $L(z,t) = a(t) + z\,\mathrm{tg}\,\alpha(t)$. The trace of the lateral surface passing in space on the time interval $t \in (t_1, t_{2N})$ forms obviously the additional part of the solid under consideration. At time moments $t \in (t_{2k-1}, t_{2k})$, $k = \overline{1,N}$, the lateral pillar surface represents the actual accretion surface of this solid. We can also consider the actual accretion surface during the continuous accretion stages as the level surface of the level t for the scalar function $\tau_*(\mathbf{r}) = \tau_*(z,\rho)$ (see Section 2):

$$\rho = L(z,t) \quad \Leftrightarrow \quad \tau_*(z,\rho) = t. \tag{5}$$

The external (relative to the accreted solid) unit normal vector to this surface can be calculated as

$$\mathbf{n}(\mathbf{r}) = \nabla \tau_*(\mathbf{r})\,|\nabla \tau_*(\mathbf{r})|^{-1} = \mathbf{e}_\rho(\varphi)\cos\alpha(\tau_*(z,\rho)) - \mathbf{e}_z \sin\alpha(\tau_*(z,\rho)). \tag{6}$$

4. The accretion problem statement and solution

Due to the objective lack of any unstressed configuration for an accreted solid the traditional kinematic description of the process of its deformation is not suitable for this solid. However, we can observe that all the particles of the added material after its attaching to the accretion surface of the solid continue to move in space as the material constituents of a solid continuum. That means that in the region of space occupied by the instantly existing accreted solid a smooth displacement velocities field $\mathbf{v}(\mathbf{r},t)$ of the particles deformation movement is uniquely determined. Therefore, the mechanical problem for the accreted solid can be stated in terms of velocities.

Let us introduce the strain velocity tensor and the so called operator stress velocity tensor

$$\mathbf{D} = (\nabla\mathbf{v} + \nabla\mathbf{v}^{\mathrm{T}})/2, \tag{7}$$

$$\mathbf{S} = \partial\mathbf{T}^{\circ}/\partial t, \tag{8}$$

respectively. Using these tensors we can rewrite material relation (2) in the velocity form

$$\mathbf{S}(\mathbf{r},t) = 2\mathbf{D}(\mathbf{r},t) + 2\nu(1-2\nu)^{-1}\mathbf{1}\mathbf{1}\bullet\bullet\,\mathbf{D}(\mathbf{r},t). \tag{9}$$

One can prove [10] that the following analogue of the standard equilibrium equation

$$\nabla\bullet\mathbf{T}(\mathbf{r},t) = \mathbf{0} \tag{10}$$

for the operator stress velocity tensor is fair in the considered case of the accreted solid deformation:

$$\nabla\bullet\mathbf{S}(\mathbf{r},t) = \mathbf{0}. \tag{11}$$

We consider the unstressed additional material being attached to the surface of the accreted pillar in the process of its accretion. So we are to put the following condition for the initial stresses at all points of the additional part of this solid:

$$\mathbf{T}(\mathbf{r},\tau_*(\mathbf{r})) = \mathbf{0}. \tag{12}$$

We can mathematically show that specific nonclassical initial condition (12) taken into account coupled with equilibrium equation (10) generates in our problem the following boundary condition for the operator stress vector velocity at points of the current accretion surface, $t \in (t_{2k-1}, t_{2k})$, $k = \overline{1,N}$:

$$\mathbf{n}(\mathbf{r})\bullet\mathbf{S}(\mathbf{r},t) = \mathbf{0}, \quad \rho = F(z,t). \tag{13}$$

It is evident that this boundary condition remains in force also in the pauses between the continuous accretion stages and after the accretion completion, because the lateral surface of the pillar is stress-free on these time segments (condition (13) has, however, a completely different mechanical nature on these time segments).

We intend to satisfy the static conditions on the expanding due to the material influx bearing end of the considered thickened pillar in integral sense. In concordance with the above formulated pillar loading conditions we have to state so the following integral boundary conditions on the bearing end surface for the stress tensor $\mathbf{T}(\mathbf{r},t) = \mathbf{T}(z,\rho,\varphi,t)$ at any time $t > t_1$ after the thickening process begins:

$$\int_0^{2\pi} \mathrm{d}\varphi \int_0^{a(t)} \mathbf{e}_z\bullet\mathbf{T}(0,\rho,\varphi,t)\,\rho\,\mathrm{d}\rho = -\mathbf{e}_z P(t), \quad \int_0^{2\pi} \mathrm{d}\varphi \int_0^{a(t)} \rho\mathbf{e}_\rho(\varphi)\times[\mathbf{e}_z\bullet\mathbf{T}(0,\rho,\varphi,t)]\,\rho\,\mathrm{d}\rho = \mathbf{0} \tag{14}$$

Notably is that it can be proven the following nontrivial corollary of these integral conditions for the velocities $\mathbf{S}(\mathbf{r},t) = \mathbf{S}(z,\rho,\varphi,t)$ of the operator stresses:

$$\begin{gathered}\int_0^{2\pi} \mathrm{d}\varphi \int_0^{a(t)} \mathbf{e}_z\bullet\mathbf{S}(0,\rho,\varphi,t)\,\rho\,\mathrm{d}\rho = -\mathbf{e}_z \frac{\mathrm{d}}{\mathrm{d}t}\left[\frac{P(t)}{G(t)} - \int_{t_0}^{t} \frac{P(\tau)}{G(\tau)} K(t,\tau)\,\mathrm{d}\tau\right], \\ \int_0^{2\pi} \mathrm{d}\varphi \int_0^{a(t)} \rho\mathbf{e}_\rho(\varphi)\times[\mathbf{e}_z\bullet\mathbf{S}(0,\rho,\varphi,t)]\,\rho\,\mathrm{d}\rho = \mathbf{0}.\end{gathered} \tag{15}$$

The proof of this corollary uses specific initial condition (12) and is tedious enough to not be placed within the present constrained paper.

Integral conditions (15) close the mathematical formulation of the stated nonclassical mechanical problem. With these conditions all the above mentioned relations give us the possibility to construct the closed-form analytical solution of the problem in question: we obtain firstly the exact in the sense of Saint-Venant principle analytical solution of the boundary value problem for differential equation (11) supplemented with relations (9) and (7), with boundary conditions (13) and (15); after that we calculate the time evolution of the operator stress tensor $\mathbf{T}^{\circ}$ at each point $\mathbf{r}$ of the ready-made pillar by integrating over time t the obtained values of velocity $\mathbf{S}$ of this tensor, with zero initial values correlated with (12); at last we reconstruct the time evolution of the true stress tensor $\mathbf{T}$ at each point by the found values of the operator stress tensor $\mathbf{T}^{\circ}$. The latter procedure can be conducted with use of formula (1) or, which is more efficient, by solving Volterra equation of the second kind

$$\frac{\mathbf{T}(\mathbf{r},t)}{G(t)} - \int_{\tau_0(\mathbf{r})}^{t} \frac{\mathbf{T}(\mathbf{r},\tau)}{G(\tau)} K(t,\tau)\mathrm{d}\tau = \mathbf{T}^{\circ}(\mathbf{r},t) \qquad (16)$$

with respect to the desired tensor-function $\mathbf{T}/G$.

The in the present study constructed analytical solution allows to retrace the evolution of the stress-strain state of the considered bearing conical pillar which is manufactured using a viscoelastic aging material under simultaneous action of an axial variable central force, after the start and on completing the piecewise continuous additive process of the pillar thickening over its lateral surface.

Acknowledgments

The present study is supported in part by the Russian Foundation for Basic Research under grants Nos. 18-01-00920-a and 17-01-00712-a.

References

[1] Manzhirov A V and Parshin D A 2015 *Mech. Solids* **50**(5) 559–70 doi: 10.3103/ S0025654415050076
[2] Manzhirov A V and Parshin D A 2015 *Mech. Solids* **50**(6) 657–75 doi: 10.3103/ S0025654415060072
[3] Manzhirov A V and Parshin D A 2016 *Mech. Solids* **51**(6) 692–700 doi: 10.3103/ S0025654416060091
[4] Bychkov P S, Kozintsev V M, Manzhirov A V and Popov A L 2017 *Mech. Solids* **52**(5) 524–9 doi: 10.3103/S0025654417050077
[5] Parshin D A 2017 *Mech. Solids* **52**(5) 530–40 doi: 10.3103/S0025654417050089
[6] Parshin D A 2018 *AIP Conf. Proc.* **1959** 070029 doi: 10.1063/1.5034704
[7] Parshin D A and Manzhirov A V 2018 *J. Phys.: Conf. Ser.* **991** 012063 doi: 10.1088/1742-6596/991/1/012063
[8] Arutyunyan N Kh 1966 *Some Problems in the Theory of Creep* (Oxford, Pergamon Press)
[9] Ross A D 1958 *J. Amer. Concr. Inst.* **29** (9) 739–58
[10] Manzhirov A V 1995 *J. Appl. Math. Mech.* **59**(5) 805–16 doi: 10.1016/0021-8928(95)00095-X

IOP Conf. Series: Materials Science and Engineering **520** (2019) 012014 doi:10.1088/1757-899X/520/1/012014

Calibration Mechanism Design and Stiffness Analysis

Siyu Xia[1], Jinfeng Shi[1], Guorui Ren[1] and Chuang Li[1,a]

[1] Xi'an Institute of Optics and Precision Mechanics, Chinese Academy of Sciences, 17 Xinxi Road.

[a] Corresponding author: lichuang@opt.ac.cn; Tel: 13991368272

Abstract. According to the optical requirements of a space camera, a calibration mechanism in the form of a worm gear driven cantilever beam is designed, and a stop is added at the cantilever beam's end for limiting its amplitude. A detailed finite element model considering the clearance between moving parts in the mechanism and the preload of bearing is established using MSC.Patran & Nastran. By comparing the vibration test and analysis results, the stiffness and the dynamic environment adaptability of the mechanism is investigated and the first-order resonant frequency of the calibration mechanism is much larger than 100Hz which is high enough for the telescope's structure. Finally, the result shows that the error between the analysis result and the test data is within 10%, and it is verified that the calibration mechanism scheme has higher stiffness.

1. Introduction

The on-orbit calibration of space cameras is the guarantee for long-term stable observation targets[1-2]. The on-track calibration of a space camera design uses the traditional built-in lamp calibration method of "LED light source + diffuse reflector". According to the optical design scheme, the diffuse reflection plate is located between the lens of the optical path and the LED lamp. As shown in Fig. 1. It needs to be turned in and out. When it is transferred into the optical path it works as a calibration. It also has the function of a shutter. Since the diffuse reflector has a small distance from the optical elements on both sides and undergoes a harsh mechanical environment during the launching process, the moving structure is required to have a high rigidity.

In the process of machining and assembly of the worm gear, the meshing backlash is inevitable, and the no-load backlash represents the inherent error of the drive combination. In the reference [3], the analysis result is higher than the test result without considering the worm gear backlash. However, the rigidity of the worm gear mechanism increases after the filler is added to eliminate the backlash, so the backlash of the worm gear has a great influence on the structural rigidity. The backlash of the worm gear can cause collision and impact between the contact surfaces of the tooth surfaces during the environmental test, thereby affecting the structural rigidity. In order to quantitatively study the influence of this contact gap, from Hertz's theory of contact problems in 1882, many practical numerical calculations of contact problems have emerged [4-5]. The formula for calculating the stiffness of angular contact ball bearings with preload is introduced in [6].

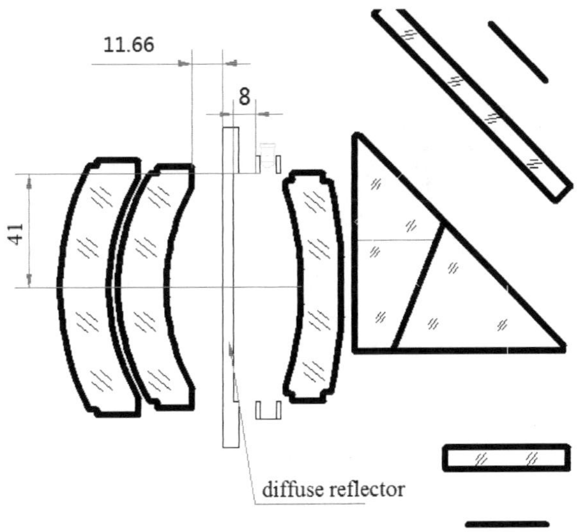

Figure 1. The relationship between optical components

The structural design of the calibration mechanism is given in this paper, and the mechanical properties of the whole mechanism are modeled and analyzed by finite element using gap theory. The structural design is improved through experiments, which effectively improves the rigidity of the mechanism. This experience provides a type of design for the future space rotation mechanism to be based on.

2. Calibration Assembly design

According to the requirements of the calibration system, the calibration mechanism needs to remain locked during launch. Due to the self-locking characteristics of the worm gear and the stall torque generated by the stepper motor, the calibration is actuated by the stepper motor and the worm gear. Because of the limited space, angular contact ball bearings are mounted back to back on both sides rather the same side of the worm wheel to increase its rigidity. In order to ensure normal operation in the space environment, all of the moving parts in the calibration mechanism are lubricated with MoS2 to avoid the phenomenon of "cold welding".

The diffuser plate of the calibration mechanism is located in the middle of the optical component with a relatively close distance, and the diameter of the calibration plate is relatively large, so that the end of the diffuser plate frame of the actuator of the calibration mechanism forms a long cantilever beam. Several designs are proposed and optimized in the reference [7] to reduce the deformation of the vibration and improve the stiffness of the system. The mechanical environment of the launch section is harsh. In order to prevent the cantilever beam from undergoing large deformation during the vibration process and colliding with the lens system on both sides, a stop limit is added at the end of the diffuse reflector frame. It also works as a hard stop for the mechanism.

The latest design of the calibration mechanism is composed of a fixed PTFE diffuse reflector of 82mm diameter, a stepper motor with 1.8° per step and a potentiometer, worm gear mechanism with a reduction ratio of 1:50, an end stop, flexible couplings avoid motion jams caused by machining and installation errors, angular contact ball bearings which can improve shaft stiffness and all necessary interfaces and housing.

The mechanism's total mass is about 1.5kg. The detailed model of the overall calibration mechanism is shown in Figure 2.

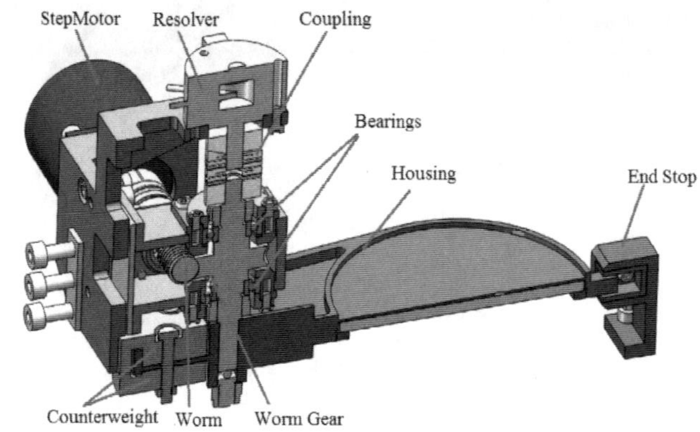

Figure 2. Detailed structure of the calibration mechanism

3. Finite Element and Modal Reduction

The calibration mechanism mainly includes worm gears, angular contact ball bearings, flexible couplings, stoppers and brackets. Different parts with different characteristics are modelled by different units to obtain results conform to the actual condition.

The solution to the contact problem requires an accurate determination of the contact state and the application of the correct contact conditions. The contact state is generally divided into three types: a separated state, a bonded contact state, and a sliding contact state. In either form, after the structure is discretized, the following contact conditions are always met between the pairs of contact points, and finally the contact conditions of the discretized form are obtained:

Normal:

$$B_n U + D_0 \geq 0 \tag{1}$$

Tangential:

$$B_t \left(U\left(t + \Delta t\right) - U\left(t\right) \right) = 0 \tag{2}$$

$U\left(t\right)$ represents the node displacement vector at time t, B_n , B_t is the contact constraint matrix, and D_0 is the initial normal gap.

The motion of any structural system can be expressed as a set of equilibrium equations for internal, external and inertial forces:

$$[M] \, \{\ddot{U}\} + [C] \, \{\dot{U}\} + [K] \, \{U\} = \{P(t)\} + \{N\} + \{Q\} \tag{3}$$

Where $[M]$: mass matrix; $[C]$: damping matrix; $[K]$: stiffness matrix; $\{P(t)\}$: external force function vector; $\{N\}$: associated nonlinear external force term vector; $\{Q\}$: boundary constraint reaction vector; $\{U\}$: displacement vector; $\{\dot{U}\}$ velocity vector; $\{\ddot{U}\}$ acceleration vector.

After the system design is completed, the mass matrix $[M]$ is constant. When the gap exists, in the state where the two objects are not in contact, the stiffness of the unit between the pair of contact points is close to zero. After reaching the area where the two objects contact, the stiffness of the unit tends to the stiffness of the contacted object, it means at this time the unit's stiffness matrix $[K]$ changes with the displacement vector $\{U\}$, and the damping matrix $[C]$ also changes accordingly, so the system becomes a nonlinear system with variable stiffness and variable damping [8-10].

The contact between the worm and the worm gear is in a separated state without external force interference, that is, the contact force is zero, and the contact surface is in a free motion state. The node pairs are selected in the contact areas of the two parts to simulate their contact stiffness using springs.

The angular contact ball bearing of the mechanism is directly connected with the structure of the worm wheel and the reduction gear box. The rigidity of the ball contact bearing has a great influence on the dynamic characteristics of the system. Under the axial preload force, the axial stiffness of the angular contact ball bearing is:

$$K_t = 6.86\left(Z^2 D_w \sin^5 \alpha_0\right)^{1/3} F_0^{1/3} \tag{4}$$

The radial stiffness is:

$$K_r = 1.7164 D_w^{1/3} Z^{2/3} \frac{\cos^2 \alpha_0}{\sin^{1/3} \alpha_0} F_0^{1/3} \tag{5}$$

In the above two formulas (4) and (5), F_0 is the axial preload (N), Z is the number of rigid balls in a single set of bearings, D_w is the diameter of the rigid ball (mm), and α_0 is the initial contact angle(°).

When the preload is too large, the torque required to drive the worm wheel is large, the wear on the motor and the coupling is large, the system's life is reduced; when the preload is too small, the system's stiffness is low, the mechanical properties cannot be met, and When the shaft system vibrates, the potentiometer will generate large sway and the accuracy will decrease. When the axial preload force is 7N, the axial stiffness is 13400N/mm, and the radial stiffness is 46780N/mm.

The flexible coupling is an important component of the transmission on the shafting system. The selection of the design has a great influence on the rigidity of the mechanism. In order to better simulate the real state, the flexible part needs to be refined as show in figure 3.

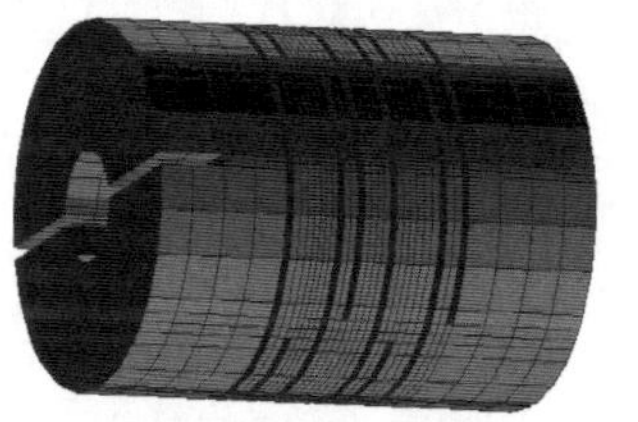

Figure 3. FEM of the Flexible coupling

For the stopper at the end of the diffuse plate frame, under the initial vibration condition, there is a certain gap between the block and the frame, and the end of the frame will collide with the polyimide on the block to generate an impact during the vibration. This contact pair is a separate contact, and the contact deformation between the two is relatively small, and is simulated by the spring unit with small stiffness.

The whole mechanism is fixed on the telescope through the four mounting points on the leftmost plane while the end stop mounted on the other side of the telescope.

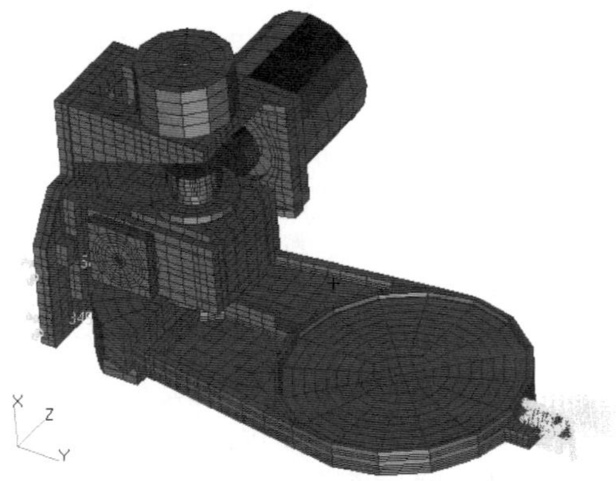

Figure 4. The mesh partition of calibration mechanism

The modeling of the structural components of the calibration mechanism mainly uses the entities (Hex8 and Wedge6). In the final model, there are 34,685 nodes and 23,070 cells. The model is shown in Figure 4. The material parameter tables used in the model are shown in Table 1 below.

Table 1. Material parameter in the model

Material	Young's modulus (GPa)	Poisson's ratio	Density (Kg/m^3)	Components
TC4	109	0.34	4430	Support
Aluminum alloy	78	0.33	2700	Worm gear, housing
45# steel	205	0.29	7850	Worm
PTFE	3	0.35	1390	Diffuse reflector
Copper	110	0.33	8930	Counterweight

4. Vibration Experiment and Analysis

In order to verify the correctness of the results of the FEM analysis and the response of the structure in the mechanical environment, the actual product was mounted on the jigs for mechanical testing. In Fig. 5, the state in which the X direction is subjected to the mechanical test is performed, and in the other two directions, it is only necessary to change the mounting direction of the tooling and the test stand.

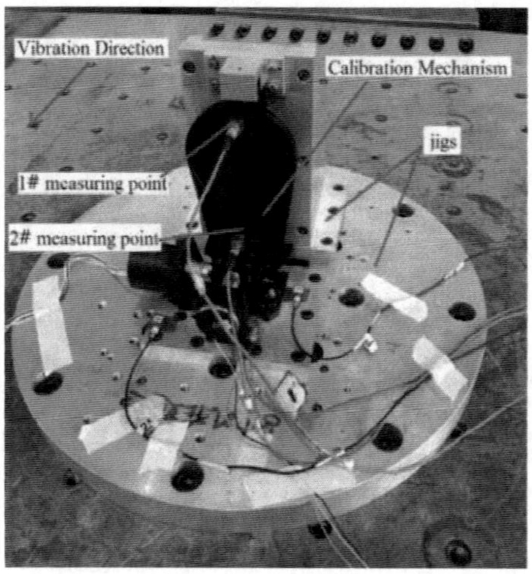

Figure 5. X-direction test

The mechanical test conditions are 0.5g characteristic sweep in 5~500Hz in 3 directions, and sinusoidal and random vibration tests in 3 directions. Through the axial low-level sine sweep test, the three-way measured fundamental frequency of the product is obtained. Table 2 gives the results of the modal analysis and the three-direction test sweep curves at the 1# measurement point.

It can be seen from Table 2 that the fundamental frequencies in the three directions are the rotation around the X, Y, and Z axes of the diffuse reflector frame, and the first-order fundamental frequency and analysis results of the characteristic sweep curves of the 1# measuring points are extracted. The fundamental frequency of the direction is summarized as shown in Table 3 below.

Table 3 shows that the fundamental frequency of the calibration mechanism is 143Hz, which is much higher than the demands of 100Hz, and the analysis results of the first-order fundamental frequency of X and Y are slightly different from the experimental results, while the results of the Z-direction first-order fundamental frequency analysis differ from the experimental results by 13 Hz, but the difference is still within 10%, and the analysis result is proven to be accurate.

Table 2. modal analysis results and experimental sweep frequency curves

	Modal analysis result	Test result (sweep curve)
X	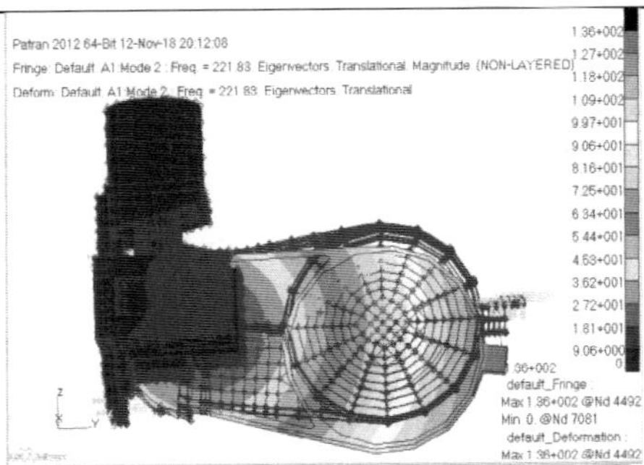	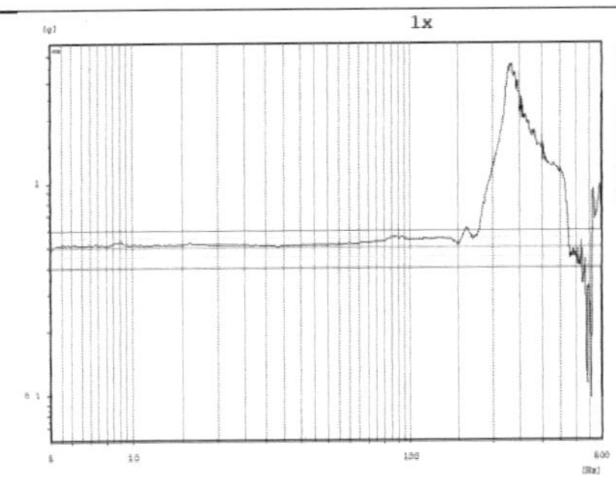
Y	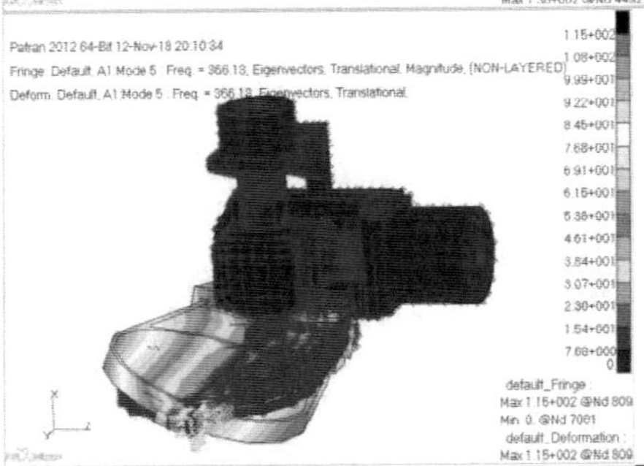	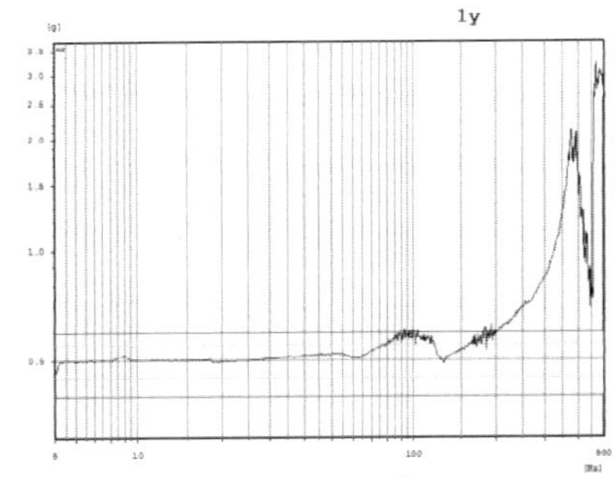
Z	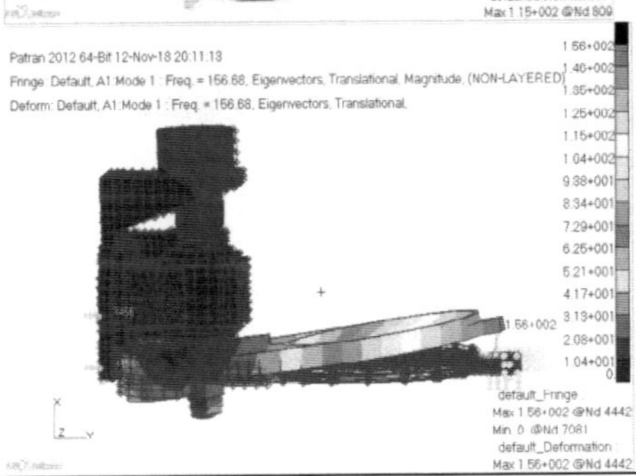	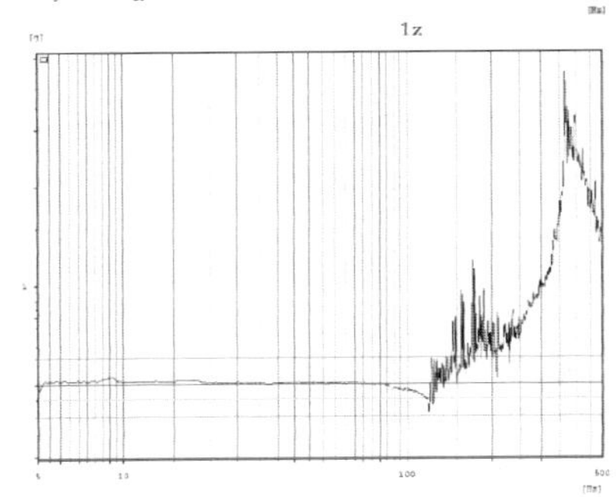

Table 3. resonant frequency

Direction	X	Y	Z
Measured (Hz)	220	376	143
Analysis result (Hz)	222	366	156
Difference (%)	0.9	2.7	9.09

5. Conclusions

According to the positional characteristics of the diffuse reflector in the optical path and the severe launching mechanical conditions, a calibration mechanism with an amplitude limit is designed. The results of finite element analysis and mechanical tests show that the first-order resonant frequency of the calibration mechanism is much larger than 100Hz, the shafting design of the calibration mechanism is reasonable and has high rigidity, which can meet the requirements of mechanical vibration. The difference between the analysis results and the test results is small, indicating that the simulation unit considering the mechanism clearance can provide a reference for the similar stiffness analysis of the rotating mechanism in the future.

IOP Conf. Series: Materials Science and Engineering **520** (2019) 012014 doi:10.1088/1757-899X/520/1/012014

References

[1] Gu X, Tian G, Yu T, Li X, Gao H, Xie Y, [Space Optical Remote Sensor Radiometric Calibration Principle and Method], Science Publishers, Beijing, 172-177 (2013)
[2] Adrian M G, Ulrich L, Alex Z and Manuel G, A Contamination Control Cover for the Mid Infrared Instrument of the James Webb Space Telescope, *Proc. SPIE 7018, 70184L1-70184L8 (2008).*
[3] W. Yongbin. Harbin Institute of Technology. (2009)
[4] H Herz, J Re. Angew Math. **92**:156-171(1882)
[5] L Shu, L Jingbo, F Ehua. J Eng Mech-ASCE, **12**:14-28(1999)
[6] Z Jingxu. Optics Precis Eng. **2**. 73-77(1996)
[7] C Yaqin, L Chuang, X Siyu, Z Peifeng, L Wang. *Proc. SPIE 10339, Pacific Rim Laser Damage 2017: Optical Materials for High-Power Lasers, 103391S*
[8] J T Stadter, R O Weiss. Computers &Structures. **10**:867-873(1979).
[9] L Zhaohui, W Zhongsu, H Qinglong, Y Liang. Chin J Sci Inst. **11**:2535-2541(2013)
[10] X Shifu, C Bin, D. Qiang, M Jun. J Dyn Control. **12**,35-40(2003)

Mathematical Model and Simulation Analysis of a New Underwater Launching System

Hao Zhou[1,a], Mao-lin Wu[1] and Xiao-er Wang[1]

[1]Naval University of Engineering Weaponry Engineering college, 430033Wuhan, China

[a] hnwzwxe@163.com

Abstract. In order to solve the problems of air turbines about complex structure, difficult processing and material requirements, a new type of vaneless underwater launch prime mover scheme is proposed. After simplification and assumption of working conditions, the launching system of the prime mover and pump is modeled and simulated by matlab. The results show that the change of chamber pressure in launcher tube is reasonable, the head and the flow of the pump is enough to provide the power for pushing the torpedo, and the other parameters such as power and speed of the prime mover scheme meet the requirements of each index, which provides theoretical support for practical application.

1. Introduction

Currently, the most advanced hydrocushion type deep-water launcher equipped on navy submarine is turbine pump launch system, which has the advantage of deep depth (600m), small volume, light weight, strong generality, high utilization ratio of launch energy and low noise, etc., making it suitable for all types of submarine[1]. Both the America and Britain have applied it to the submarine of 1990s. For the rotary pump launch system, other aerodynamic device could also be used as prime motor to drive the pump despite the air turbine[2]. Air jet rotary engine is one choice of high feasibility. This paper establishes the mathematic model of air jet rotary engine on the foundation of introducing the structure, operating principle of it, and then builds the simulation software to study the launch process of this launch system.

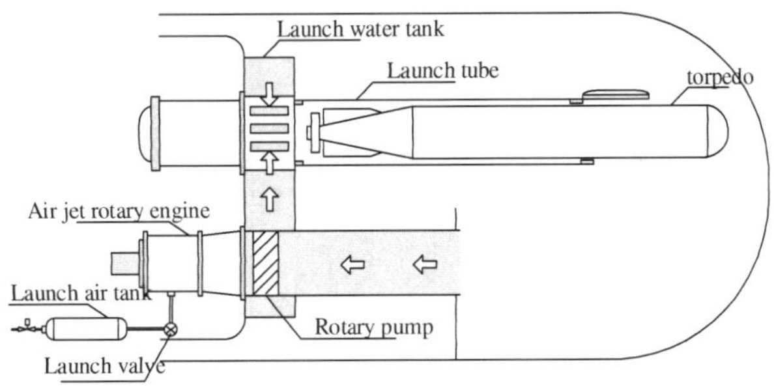

Figure.1 Schematic of AJRP launch system

2. The structure and basic operation principle of AJRP

AJRP lunch system mainly consists of launch air bottle, launch valve, air jet rotary engine, rotary pump and launch tube, as is shown in Fig.1. The operation principle of AJRP is: when launching, the valve opens at a certain rule, then the high pressure gas goes into air jet rotary engine through the valve and drive the engine, and the engine drives the pump through moderator, at last the pump pumps the water into launch tube and launch the weapon.

As is shown in figure.2, air jet rotary engine is a rotary engine device with high pressure air as working medium, which takes advantage of the counter-acting force of high speed jet airstream to generate rotating torque and power. It consists of shell, air jet wheel and moderator, and the output shaft of moderator is connected to the load.

3. Basic assumption of launch system modeling

The launch process of real launch system is really complicated. To make the problem easier to study, assumption is made that: the phase of air flows through launch valve is a throttling process of adiabatic expansion; the decompression throttling process of air flows in the air jet engine pipe is adiabatic process; the wear-stress of weapon in launch pipe is under simplified analysis, which ignores the elements like additional mass, which have small influence on the weapon movement; the water in the launch pipe is considered to only make work on weapon, ignoring the working to the out flowing water through the interval of weapon and pipe.

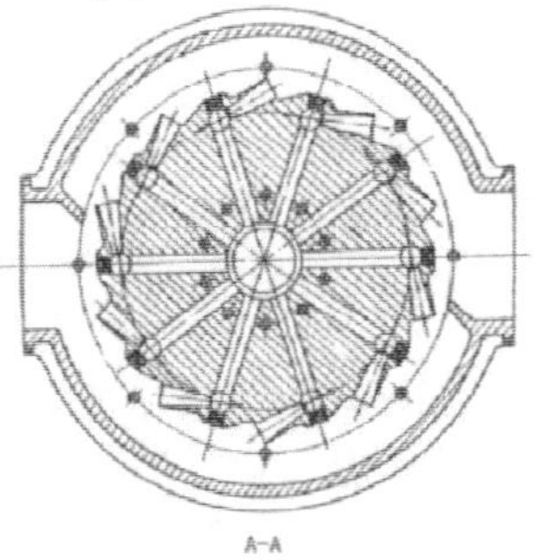

(a) A-A profile

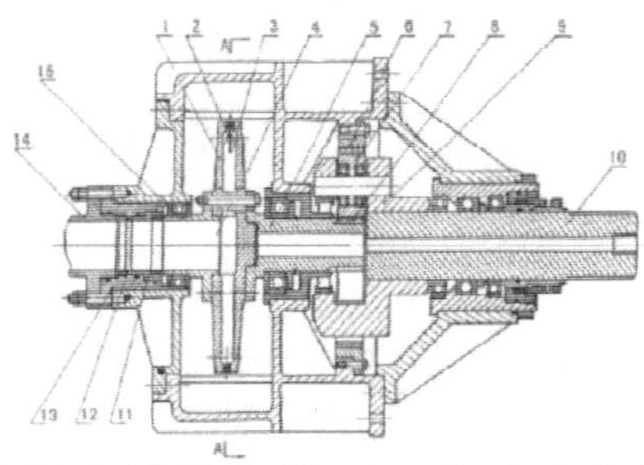

1-air jet wheel, 2-jet pipe, 3-branch air flue, 4-main air flue, 5-transmission shaft, 6-internal gear, 7-planetary gear, 8-central gear, 9-planetary gear carrier, 10-output shaft, 11-mating ring, 12-spring, 13-adjust volution, 14-connecting pipe, 15-intake tube

(b) Axial profile

Figure.2 Schematic of Air-jet Rotary

4. Modeling of launch system

4.1 Mathematic model of launch gas tank gas state

This model is used to describe the variation of gas pressure, temperature and mass during the launch process.

Assuming that the gas tank air-supply process during the launching is an adiabatic process, then through the law of thermodynamics for gas we can get:

$$\frac{\mathrm{d}p_{\mathrm{B}}}{\mathrm{d}t}=\kappa\frac{p_{\mathrm{B}}}{m_{\mathrm{B}}}\frac{\mathrm{d}m_{\mathrm{B}}}{\mathrm{d}t} \tag{1}$$

$$\frac{\mathrm{d}T_{\mathrm{B}}}{\mathrm{d}t}=(\kappa-1)\frac{T_{\mathrm{B}}}{m_{\mathrm{B}}}\frac{\mathrm{d}m_{\mathrm{B}}}{\mathrm{d}t} \tag{2}$$

In which p_{B} is the air pressure in the launch gas tank, T_{B} is the temperature of the tank, m_{B} is the mass of the tank, κ is the adiabatic index.

4.2 Mathematic model of launch valve

The amount of gas flow in and out of the engine is adjusted by the launch valve, which is realized through the opening area changing with the time. In the simulation the changing rule is set to $S_{\mathrm{v}}=f(t)$.

4.3 Mathematic model of air jet rotary engine

The model of air jet rotary engine includes three parts: the model of jet pipe, calculation model of rotate torque of pipe utter and kinematic model of air jet rotary engine.

4.3.1 Mathematic model of jet pipe This model mainly solves the gas flow speed of pipe utter.Assuming that the flow of the pipe is isentropic adiabatic process, obeying the flow basic equations with one unknown[3], energy conservation equation can be obtained through equation (3)-(5):

$$c_{\mathrm{p}}T_{\mathrm{no}}+\frac{w_{\mathrm{no}}{}^{2}}{2}-\frac{u_{\mathrm{no}}{}^{2}}{2}=c_{\mathrm{p}}T_{0} \tag{3}$$

T_{no}, w_{no} and u_{no} represent the gas temperature, relative speed and circular velocity of pipe utter section respectively, T_0 is relative stagnation temperature. Assuming that high pressure gas flows through valve without loss and the process being adiabatic, generally $T_0=T_{\mathrm{B}}$, T_{B} is the temperature of launch gas tank.

Flow conservation equation of pipe flow process is as follow:

$$q_{\mathrm{nm}}=q_{\mathrm{no}} \tag{4}$$

In which q_{nm} is the flow of smallest section of pipe, q_{no} is the flow of utter section.

On the smallest section, the pressure is critical pressure ratio, and then the flow q_{nm} is:

$$q_{\mathrm{nm}}=\varphi_{\mathrm{n}}S_{\mathrm{nm}}\varepsilon_1\bullet\frac{p_{\mathrm{ni}}}{\sqrt{RT_{\mathrm{B}}}} \tag{5}$$

In which φ_{n} is the flow coefficient of pipe, p_{ni} is the pressure of the entrance of pipe, S_{nm} is the area of the smallest section of pipe, R is the air gas constant. $\varepsilon_1=\left(\frac{2}{\kappa+1}\right)^{\frac{1}{\kappa-1}}\cdot\sqrt{\frac{2\kappa}{\kappa+1}}$

The flow q_{no} of utter section can be calculated as follow:

$$q_{\mathrm{no}}=\varphi_{\mathrm{no}}S_{\mathrm{no}}\rho_{\mathrm{no}}w_{\mathrm{no}} \tag{6}$$

In which φ_{no} is the flow coefficient of utter section, S_{no} is the area of utter section, ρ_{no} is the air density of utter section, w_{no} is the flow velocity of utter section.

With gas adiabatic equation[4] we can get that:

$$\frac{\rho_{\mathrm{no}}}{\rho_0}=\left(\frac{T_{\mathrm{no}}}{T_0}\right)^{\frac{1}{\kappa-1}} \tag{7}$$

$$\frac{p_{no}}{p_0} = \left(\frac{T_{no}}{T_0}\right)^{\frac{\kappa}{\kappa-1}} \tag{8}$$

In which ρ_0, p_0 is the air density and pressure when the temperature is T_0.

Substitute equation (5), (6) into (4) and combine them with (7) we can get that:

$$T_{no} = \left(\frac{A_{n1}}{w_{no}}\right)^{\kappa-1} \tag{9}$$

$$A_{n1} = \varphi_n S_{nm} \varepsilon_1 \bullet \frac{p_{ni}}{\varphi_{no} S_{no} \rho_0 \sqrt{RT_B}} T_0^{\frac{1}{\kappa-1}} \tag{10}$$

Substitute equation (9) into (5) and solve the equation with numerical methods, then the relative velocity of air in pipe utter section when the air jet wheel speed is known[4].

4.3.2 Calculation model of air jet rotary engine rotate torque To calculate the torque and power when the engine is running, single pipe should be analyzed firstly (as is shown in Fig.3).

The rotate torque produced by the work of single jet pipe is made up of two parts: the moment of momentum produced by gas ejection and the moment produced by the pressure differential of the ejector.

(1) Calculation of the moment of momentum produced by gas ejection

When the air flows through the center pipe, the entrance velocity is $\mathbf{v_1}$, the relative velocity of pipe utter section is $\mathbf{w_{no}}$, which keeps an angle of β with the direction of wheel tangent. Assuming that the flow in the pipe is isoenergetic isentropic flow. Take the gas in AABB as the object, after the time of dt ,it moves to the position A'A'B'B', then AAA'A' is mass of the part that flows into pipe, marked as dm_1 ; **BBB'B'** is the mass of the part that flows out of the pipe, marked as dm_2 . Assuming that the parameter of A'A'BB during dt remain unchanged, then:

$dm_1 = dm_2$

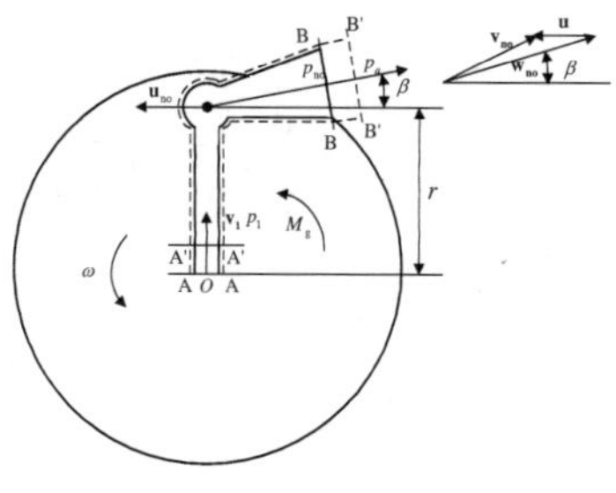

Figure.3 Schematic of jet pipe

Take point O as the rotate center, then according to the theorem of moment of momentum:

$$M_{T1}dt = dm_2 \cdot \mathbf{r}_2 \times (\mathbf{w_{no}} + \mathbf{u_{no}}) - dm_1 \cdot \mathbf{r}_1 \times \mathbf{v_1} \tag{11}$$

M_{T1} is the joint momental vector of the air mass; $\mathbf{r}_1$ is the radius vector of entrance; $\mathbf{r}_2$ is the radius vector of the utter; $\mathbf{u_{no}}$ is the peripheral speed of the utter.

Assuming that anticlockwise on the plane where air jet wheel lays is positive, since $\mathbf{r}_1 = 0$ and the module of $\mathbf{r}_2$ is $r_2 = r$, and $dm_1 = dm_2 = dm$, then the scalar quantity after trimming is as follow:

$$M_{T1} = r(u_{no} - w_{no}\cos\beta)\frac{dm}{dt} = r(u_{no} - w_{no}\cos\beta)\dot{m} \tag{12}$$

In which $\dot{m}$ is the air mass flow of single pipe, the peripheral speed of pipe utter is $u_{no} = \omega\, r = 2\pi n_{tur} r$, and ω is the air wheel angular velocity, n_{jet} is the air wheel rotating velocity.

Since the direction of the joint moment on the air mass in opposite to that of moment it cast on the wheel and the size of two moment is the same, the kinetic moment produced of single pipe ejection is: $M_{g1} = -M_{T1}$

(2) The moment produced by pressure differential of utter

The utter of pipe is connected to the outside, so the pressure moment on the air jet wheel is imbalanced. Assuming that the outside air pressure is p_a, the air pressure of utter is p_{no}, then the moment produced by pressure differential of utter is:

$$\begin{aligned} M_{g2} &= rp_{no}A_2\cos\beta - rp_a A_2\cos\beta \\ &= (p_{no} - p_a)rA_2\cos\beta \end{aligned} \tag{13}$$

(3) The joint momentum on the air wheel

When the number of pipe is n_j, according to the analysis above, the joint momentum is:

$$M_{jet} = n_j(M_{g1} + M_{g2}) \tag{14}$$

4.3.3 The motion model of air jet rotary engine The motion equation of the procession of the engine driving rotary pump[5-6] is shown as:

$$\frac{\pi}{30} I_{jet} \frac{dn_{jet}}{dt} = M_{jet} - i_r\left(M_p + \frac{\pi}{30} I_p \frac{dn_p}{dt}\right) \tag{15}$$

In which the rotational inertia of rotation pump is I_p, that of the air rotary engine is I_{jet}, the torque of rotation pump is M_p, and the initiative torque of air jet rotary engine is M_{jet}, and the rotation velocity of rotation pump is n_p, that of the air jet rotary engine is n_{jet}, and the reduction ratio of the moderator is $i_r = n_p / n_{jet}$.

After summing up, the angular accelerated velocity of air wheel is:

$$\frac{dn_{jet}}{dt} = \frac{30(M_{jet} - i_r M_p)}{\pi(I_{jet} + i_r^2 \bullet I_p)} \tag{16}$$

After solving the equation (14) the impeller rotation velocity n_{jet} is obtained, thus the periphery velocity u is:

$$u = \frac{\pi}{30} r \bullet n_{jet} \tag{17}$$

r is the radius of the pipe utter.

The output torque and rotation velocity of air jet rotary engine is known, and then the output power is:

$$N_{jet} = \frac{\pi}{30} M_{jet} n_{jet} \tag{18}$$

4.4 Model of rotation pump

The mathematic model of rotation pump is main to describe the relationship of pump and the parameters like output lift, flow and power, etc.

Combining with the lift curve of pump, the equation of lift curve of pump under any rotation velocity is:

$$\begin{aligned} H_P &= H(n_P, Q_P) \\ &= 6.1583\times10^{-5} n_P^2 + 5.217\times10^{-4} Q_P n_P - 3.9660 Q_P^2 \end{aligned} \tag{19}$$

Q_P is the flow of pump.

The life is known, and the drag torque of pump to the air jet rotary engine principal axis is:

$$M_p = \frac{30 N_{pe}}{\pi n_P} \tag{20}$$

In which N_{Pe} is the valid power of pump, $N_{pe} = \eta_P p_o Q_P$, η_P is the power of pump, p_o is the output water pressure of pump, $p_o = 1.02\times10^4 H_P$ (Pa).

4.5 Kinematic model of weapon launch tube

During the launch process, the force cast on the weapon in the launch tube is(as is shown in Fig.4): gravity G , buoyancy F , head resistance $A_x(v_T + v_{sub})^2$, friction f_g between the weapon and the tube, water pressure p_t of weapon stern and the external water pressure p_H ,the motion equation of which is[7]:

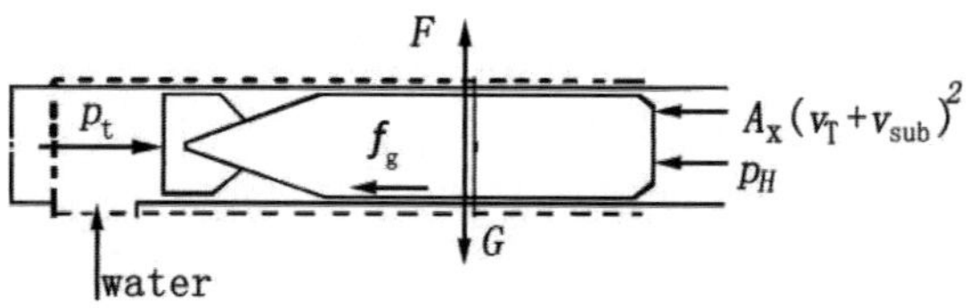

Figure.4 Schematic of the force acting on the torpedo in tube

$$\frac{dv_T}{dt} = a_T = \frac{1}{m}\left[p_c S_T - f_g - A_x(v_T + v_{sub})^2 - S_T \rho v_T^2\right] \tag{21}$$

$$\frac{dl}{dt} = v_T \tag{22}$$

$$m = m_W + m_T = m_{W0} + \rho S_T l + m_T \tag{23}$$

$$p_c = p_t - p_H;\ p_t = p_o - \Delta p \tag{24}$$

S_T is the area of weapon section; v_T , a_T is the velocity and the acceleration velocity of weapon; l is weapon displacement inside the tube; p_c is the chamber pressure of the launch tube; Δp is the pressure loss during the pump to the tube; m_T is the mass of weapon; m_W the mass of water that move with the weapon; m_{W0} is the initial mass of the water moves with the weapon. $f_g = \mu(G - F)$, μ is the friction coefficient; v_{sub} the voyage velocity when the weapon is launched; A_x is the head on fluid resistance coefficient; ρ is the water density.

5. Calculation simulation of launch process and the analysis of simulation result

After building simulation software according to the models above, simulation can be conducted to the process of AJRP launching weapon.

IOP Conf. Series: Materials Science and Engineering **520** (2019) 012015 doi:10.1088/1757-899X/520/1/012015

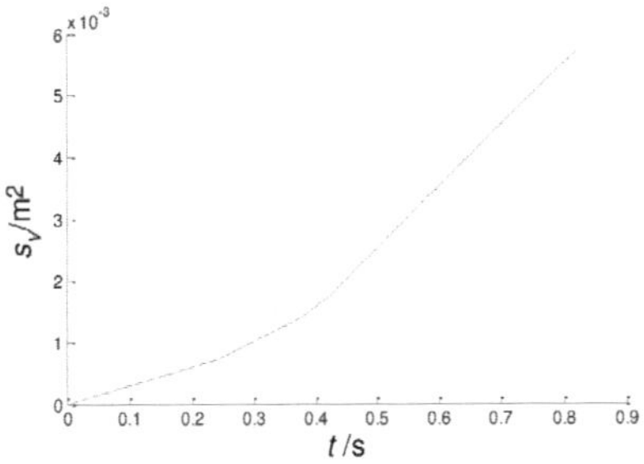

Figure.5 The area variation of launching valve

5.1 main parameters of simulation

Initial pressure of gas tank:15MPa; initial temperature :300.0K; volume :300L; torpedo CSA: 0.2254m^2; weight: 1850.0kg; water density: 1040.0kg/m^3; displacement: 6.0m; head on fluid resistance coefficient :16N.s^2/m^2;air wheel rotational inertia I_t : 1.0 kg.m^2; air wheel pipe utter: 0.25m; number of pipe :9, position angle of pipe utter r: 0.25m;reduction ratio of moderator ;pump blade rotational inertial :13.0kg.m^2; voyage velocity when the torpedo is launched: 5m/s.

The changing law of launching valve opening area is shown in Fig.5.

5.2 Simulation result

The simulation result of air jet rotary engine is shown as follows.

(1) Simulation result of basic parameters

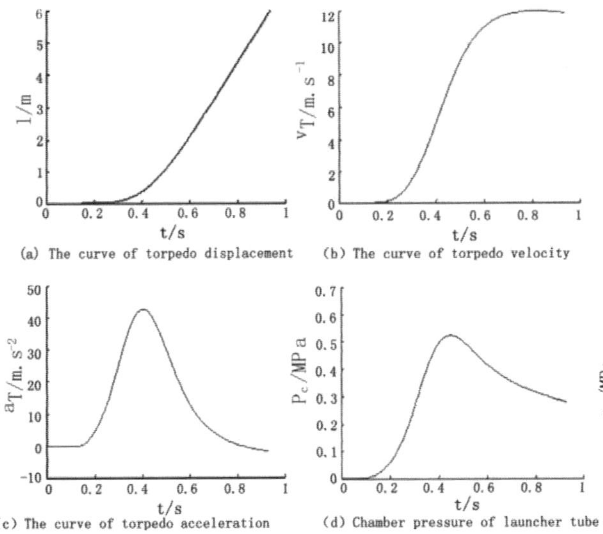

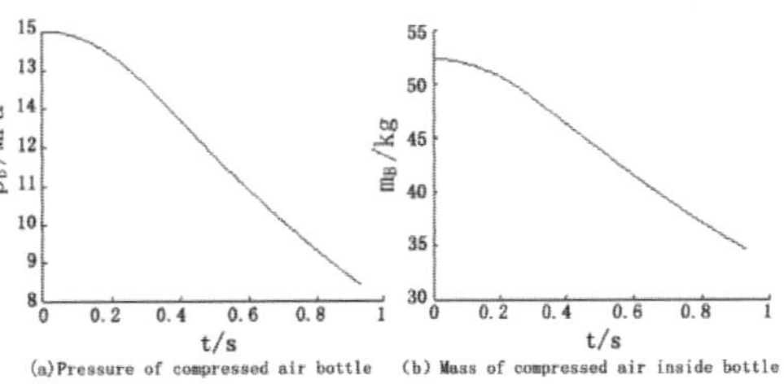

Figure.6 Simulation results of turbine-pump launch system **Figure.7** Operating condition of bottle

5.3 Analysis of simulation result

It can be learnt from the simulation result that: with fixed parameters, AJRP can push the torpedo to the utter in 0.9s (as is shown in Fig.6a), of which the speed hits 12.0m/s (as is shown in Fig.6b), the max acceleration velocity of torpedo is 42.6m/s^2 (as is shown in Fig.6c), which meets the demand of submarine for velocity and acceleration velocity. Launching chamber pressure is 0.52MPa (as is shown in Fig.6d), which meets the demand of security.

The required air power can be obtained through simulation. When the launch is over, launch bottle pressure of AJRP has reduced to 8.4MPa from 15MPa, the mass of gas has reduced to 34.56kg from 52.26kg, the total loss of gas is 17.70kg (as is shown in Fig.7).

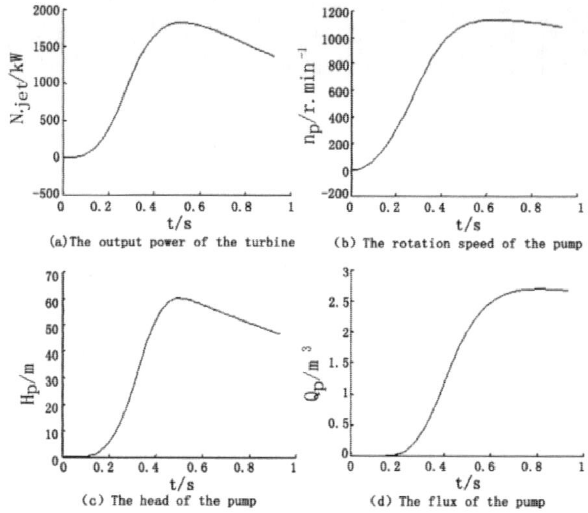

Figure.8 Dynamic characteristics of turbine and pump

The dynamic character of air jet rotary engine can also be obtained. Air jet rotary engine can provide the maximum power of 1823KW (as is shown in Fig.8a), when the rotation speed of pump is increasing to 1130r/min from 0r/min (as is shown in Fig.8b), the head increases to 60m/s from 0r/s (as is shown in Fig.8c), and the flow of pump is increasing to 2.7 m^3/s (as is shown in Fig.8d). From the simulation result, the character of air jet rotary engine meets the demand of AJRP launch system. According to the simulation model, the influences of main parameters of launch system like rotational inertia, size of flow area and initial pressure of bottle to the system can also be studied.

6. Conclusion

The simulation result of AJRP launch system shows that, the established model for this system is basically correct. The simulation software based on these models can not only be used to simulate the system and study its character, but can also be used to provide theory foundation for the design of the system and the key module like air jet rotary engine, launch valve and rotation pump.

References

[1] Lian Yongqing, Wang Shuzhong. Principle of torpedo launcher [M].Beijing:National Defence Industry Press,2012.4:110-115

[2] Baoguo Wang, Shuyan Liu. Gas Dynamics[M]. Beijing: Beijing Institute of Technology Press, 2005.8:104~106

[3] Xiaofang Zhang, Shuzong Wang, Yongqin Lian. Modeling and Simulation of Pneumatic and Hydraulic Underwater Weapon Launching System[J]. Journal of System Simulation, 2009, 21(10): 3092-3095

[4] HUA Bin-bin, WANG Rui-lin, MA Long, et al. Dynamics Modeling and Simulation of Self-Powered Gatling Gun[J]. Journal of Ordnance Engineering, 2017.2: 23-27.

[5] Juraeva Makhsuda, Ryu K et al. Optimum Design of a Saw-Tooth-Shaped Dental Air-turbine Using Design of Experiment, International Journal of Precision Engineering and Manufacturing, 2014,15(2):227-234.

[6] Juraeva Makhsuda, Bong Hwan Park, Kyung Jin Ryu, et al. Designing High-speed Dental Air-tube Handpiece by Using a Computational Approach[J]. International Journal of Precision Engineering and Manufacturing, 2017.10:1403-1407.

[7] NIU Qing-yong, LI Tian-yun. Numerical Simulation on the Internal Flow Field of an Underwater Pneumatic Launcher Model[J]. Machine tool & Hydraulics, 2015, 43(4): 58-61.

Research on Immersion of Interactive Image Design in Visual Communication

Jinpeng Xu

Beihang University, 37 Xueyuan Road, Haidian District, 100191 Beijing, China

E-mail: 1683281736@qq.com

Abstract. In the course of new media art development, immersion is an ineluctable topic. Under the context of design, interactive image features immersion. The role of immersive design is not to add a new property, but to enhance existing properties. Virtual interface and individual interaction, as two main characteristics of immersion in visual communication, testify it on the other side. The paper discusses and analyzes narrative modes of visual immersion and structure of immersion in works from the perspective of visual language, and concentrates on non-structure design language adopted to organize visual information to develop a new road for experience design languages.

1. Introduction

Immersion is a description of a technology, and describes the extent to which the computer displays are capable of delivering an inclusive, extensive, surrounding and vivid illusion of reality to the sense of a human participant. This is the definition of immersion presented by Mel Slater and Sylvia Wilbur in 1997 [1]. According to this definition, immersion is not a kind of psychological process, but a kind of physical process, which could be interpreted from the perspective of technology. During this process, our body and feelings, cheated by virtual scenes, will generate related actions and responses accordingly. Many researchers at the present times such as Wijnand Ijsselsteijn [2], Giuseppe Riva [2] and Rettie [3] follow this definition; but also admit immersion is a kind of psychological process, and supplement the definition of immersion [4]. The paper will also follow definition of immersion presented by Slater and Wilbur and deem it as the objective attribute.

Under current context of design, boundary of visual communication is expanded constantly, and visual communication stresses more on the bidirectional. Higher requirement, i.e. immersion, is presented by viewers on interactive image design, whose properties are marginal in traditional graphic design. Even now, it exists as a kind of non-structure design language. Without it, a good work could be created, too. However, immersion could create a special sense judgment to touch viewers; it will also fill a work with poetry, and make such work have the power of communication [5].

2. Characteristics of immersion on visual communication design

Function and relationship should be stressed when discuss immersion under the context of design. Its characteristics are also developed based on these two points. Summarizing and analyzing some works, characteristics of immersion on visual communication design such as virtual interface and individual interaction could be found. The next part will state it based on Archive Dreaming (2017) created by

Refik Anadol[1] and ANIMA IKI (2015) created by Onformative[2]. Meanwhile, the role of immersion is to enhance the existing communication effect of work, but not to impose a property to it. Both works were created by taking interactive image as the core of design and stressing expression of existing immersion at the same time.

Refik Anado, a new media artist and director born in Istanbul, now works and live in Los Angeles. Archive Dreaming figure 1, exhibited in SALT Galata Art Museum in Istanbul, is a shocking data visualization work co-created by Refik Anadol and Google. With 1.7 million documents owned by the art museum as the data source, the work shows 1.7 million pieces images in real time through special programming. With the help of data visualization design system, enormous data files are transformed into immersive space, which embraces audiences. Meanwhile, the system also shows information of each image, and audiences could interact with such visual materials in real time.

Figure 1. Archive Dreaming.

Virtual interface is generated on the walls through projection; and interactive operation is done with the help of tablet personal computer. Mirrors are used on ceiling and floor to reflect images on the walls. Virtual interface generated in this way has two functions. The first one: a large screen is needed to show so many works clearly; however, the space is limited; therefore, virtual screen is the best choice. The second one: it offers better condition for audiences to watch such process, then to reach focused situation. At the same time, such effect is heart-shaking, that audiences' desire of experience is stimulated greatly, which is beneficial to gain contents.

Design of interactive images could be shown on mobile phone as well as in a space. There is no denying that it still features immersion even it is shown on small screens like mobile phone screen, and people are attracted as before. As its modes of presentation and technologies are changed and innovated constantly, immersion is enhanced continuously. Audiences' experience is optimized when functions of presentation and watch are strengthened. This is more than the changes happening when it is moved from a small plane to a space. It is improvement on function and users' satisfaction which drive it to change. Designers give enough thought to the importance to express immersion.

The other case is from Onformative, a design and digital art studio in Berlin. ANIMA IKI figure 2, its image is actual a gigantic sphere, is known by people because it won the Fifth Lumen Prize. It creates immersive and captivating watch and interaction experience, and explores relationship between the image and its surroundings through sport, texture, light and sound. The sphere, the only light source in the space, shines, drawing tourists to visit; it reacts to the appearance of tourists. People

[1] Anadol R 2017 *Archive Dreaming* Available from: https://vimeo.com/218573298
[2] Onformative 2015 *ANIMA IKI* Available from: https://vimeo.com/128767230

could find that sticky metal fluid flows constantly on the surface of the sphere, which distorts the surface and creates a kind of fascinating mysterious scene.

Figure 2. ANIMA IKI.

Interactive image is connected to the individual naturally. Such connection is generated by interactive action and image. Visual design relies on more than a kind of instrumental language to build a receptive relationship; at the same time, it pays more attention to the non-structure design language adopted to organize visual information to develop a road for experience design language. Therefore, individual interaction becomes an obvious characteristic of immersion under the context of visual communication design [6]. Link degree of immersion with users' experience. Individual interaction makes connection between work and users become closer; this is specific analysis on relationship. The stronger the interaction of the image, the better the visitors' experience; immersion could reach higher degree, or even enter the status of Deep Immersion. On the contrary, the worse the interaction, the worse the visitors' experience, the lower the degree of immersion; part of properties of immersion will be ignored by visitors that they could only accept part of information the designer wants to pass on.

Design of ANIMA IKI is simple, but is able to arouse people's inmost reverie, and adopts individual interaction directly. Getting close to the sphere, visitors' actions will be directly reflected on the flow of sticky metal fluid. This makes interaction become easy and interesting. Therefore, individual interaction is testified that it is an extremely effective tool other than the characteristic of immersion in design. Employ it a little to create the work, then it will greatly reduce the distance between audiences and the work.

Oliver Grau describes the manifestation mode of immersion on Virtual Art from Illusion to Immersion (2003): the majority of virtual realities that are experienced almost wholly visually seal off the observer hermetically from external visual impressions, appeal to him or her with plastic objects, expand perspective of real space into illusion space, observe scale and color correspondence, and like the panorama, use indirect light effects to make the image appear as the source of the real [7]. This is immersive design at the beginning of 21st century. As technologies of screen display and projection develops constantly, plastic objects are replaced by interactive images; only rules about proportion, color and light are kept. In 2007, Alex McDowell presented concept of immersive design for the first time. In 2008, conference of 5D - The Future of Immersive Design[3] that he participated in holding exerts an important influence on professional field. Under the background of contemporary design, interactive images replace traditional still pictures and become a new manifestation mode. Although both of them belong to visual communication design, the whole research context changes.

[3] *5D- The Future of Immersive Design* Available from: https://www.awn.com/event/5d-future-immersive-design

IOP Conf. Series: Materials Science and Engineering 520 (2019) 012016 doi:10.1088/1757-899X/520/1/012016

According to analysis on characteristics of these two works, it could be found that immersion could not be deemed as a kind of design philosophy. Because each image has such intrinsic attribute. Degree of immersion depends on audiences' acceptance level on image. The higher the audiences' acceptance level, the stronger the intrinsic attribute. What immersive design does is to enhance the expression of the design; and methods to realize such effect are diverse that it is not limited to specific method. Space, projection, equipment or even auditory sense and smell (Discussion on the paper will not involve auditory sense and smell, but just focus on visual immersion.) are worthy of consideration.

3. Nonlinear narration of visual immersion

Nonlinear narration is frequently employed in literature and films. Montage, as a kind of film language, is nonlinear to some extent. Under the context of visual communication, different from linear narrative logic of traditional design, interactive image is presented with a kind of separated, discrete and free form, that design is done by breaking conventional images, characters and grids. It is a kind of nonlinear narration when it is shown to audiences [8]. Specific features of manifestation contain various visual angles, contingency and fragmentation.

3.1. Various visual angles

The generation of multi-perspective narration of visual immersion is related to forms as well as clues used to convey information. Visual immersion has relation to space. The space, as a kind of method, enhances the immersion. Interactive images have more varied expression forms within the space, which will generate more clues. The objects' direction, location or sequence are rid of physical laws; geometrical morphology, scenography and displacement also gain brand-new conditions in abstract space. It is inevitable for visual immersion in spatial relationship to generate multi-perspective nonlinear narration. It could be illustrated by works such as Floating, Resonating Spheres, Living Digital Forest and Future Park created by Team Lab, a design team in Japan. Designers design several main lines for images existing in the space, therefore it forms multi-perspective nonlinear narration mode. Meanwhile, such design is common that directly employ multiple screens (or multiple channels) to narrate respectively.

Multi-perspective narration mode is employed to enhance visual immersion, then to reach deep immersion. Information will be easily accepted by audiences and difficulty to pass on information will be reduced with the help of multi-perspective information transfer. Human brain will process information again to get close to the core of design step by step. Combination of information transfer which seems normal from single perspective will arouse audiences' feelings; it is significant.

3.2. Contingency

Dada has a great influence on visual communication design; the important point is that it confirms the contingency and opportunity of design. Different from some classical design works, its narrative mode is relatively closed; it pursues for relationship between elements and fixed final presentation. Interactive design could maintain ideas and forms generated during the process and show these ideas and forms on the final result. This makes it become inclusive and open; there is no changeless form. In visual immersion, such contingency is expanded because of audiences' participation. Individual difference will lead to different results. Contingency of narration increases greatly, becoming the obvious narration mode of visual immersion. It is an important part of work.

3.3. Fragmentation

Various visual angles, multiple clues and contingency inevitably create fragmentation for the whole structure. Image is presented to audiences with scattered form, that audiences should piece them together with their own consciousness when receive information to form completed narration. This is a new visual communication mode. However, problems are as plain as the nose on your face. Such combination is equivalent to re-creation. Different people has different experience, knowledge reserve and value; furthermore, information may miss during information transmission; all of these lead to deviation on veracity, which should be amended through characters. Fragmentation itself becomes the biggest attraction of this narration; it will also make design become interesting, and enhance visual immersion. Such sense of fragmentation brought by the narration becomes one of aesthetic tastes of design.

4. Structure of immersion in interactive image

As a kind of nonlinear narration method, structure of visual immersion is decided by designer. The main part of design is interactive image, which is also affected by audiences' interaction. How to guide and allow audiences to build design skeleton and narration structure through interactive methods are the primary issues which should be solved at the conceptual phase of design [9, 10]. There are many structures and models of immersion design now; however, there are mainly three structural models after summarizing.

4.1. Dot structure

This is the simplest structural model used to do immersion design; it is widely applied to interactive image. Take Interaction Show of Atec (hold by Alibaba)[4], the interactive show link done in the middle of conference, for example figure 3. Interactive image in the center would change as on-the-spot sound changes when all lights are turned off. Similar to ANIMA iki, audiences concentrate on the middle big screen; and change of interactive image would affect itself only, but would not generate relation with other items. This is a kind of structure that is easy to be accepted by designer and audiences; its form is extremely concise, but it could not do anything for complicated exhibition. Under the totally dark environment, the only light source changes slowly, and people are attracted by it unconsciously. It is easy for people to reach immersion status. Therefore, this is the simplest and directest design structure which could give full play to the immersion of interactive image.

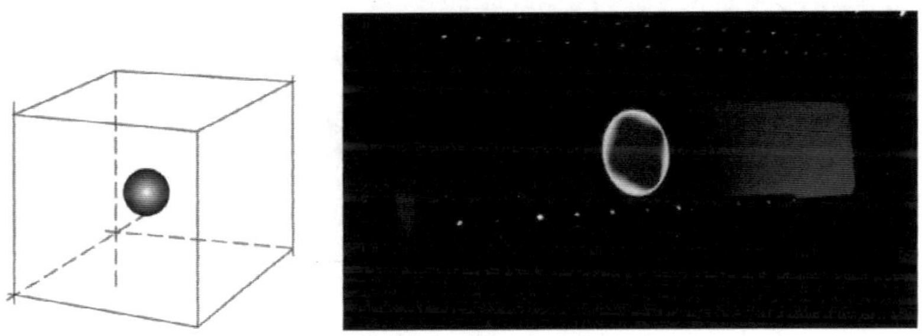

Figure 3. Model & Example.

4.2. Chain structure

Composed of many dots, individual chain structure could exist independently, but there is a clue or theme which links them together. For this structure, connection between independent units is unidirectional. It reaches richer and more shocked expression. Even interaction affects some of units, the whole chain will not change. This is similar to parallel montage, that audiences participate in, but they will not affect the results, because final results are the fixed forms or purposes which have been designed in advance. Such structure could operate on part of image, or on whole image. It will not affect users' experience while it enhances audiences' immersion.

Take Atom created by Christopher Bauder[5] for example figure 4. This is a dynamic presentation sculpture which is chain matrix structure. Interactive light images are controlled by sounds. Each light is relatively independent, that they have their own setting and respond to different sounds. Computer is

[4] *Interaction Show of Atec 2017* Available from: https://tech.antfin.com/articles/activity/atec

[5] *Atom* Available from: http://www.christopherbauder.com/

employed to control cable to adjust the height of balloons. Each dot is equivalent to an independent pixel; all of them form the image.

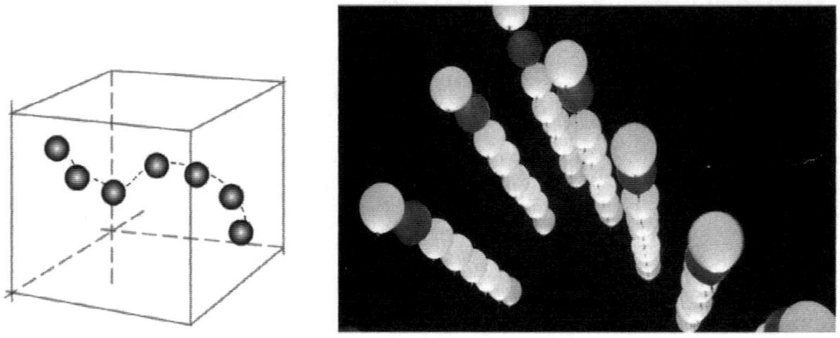

Figure 4. Model&Example.

4.3. Net structure
This structure conforms to our daily life scenes, which are fragmentation and linked together with several main lines. To learn a person, we may learn his work at first, then stories hidden behind his work, then learn his student-days, and finally learn his family status. Lots of events and people related to him will be added in this process. This forms a web map. It will be easy for audiences to accept if designer well arranges the internal logic. This net structure is extremely suitable for bearing huge data, and to present data visualization. It will reach excellent presentation effect while save cost. Take Archive Dreaming created by Refik Anadol for example figure 5.

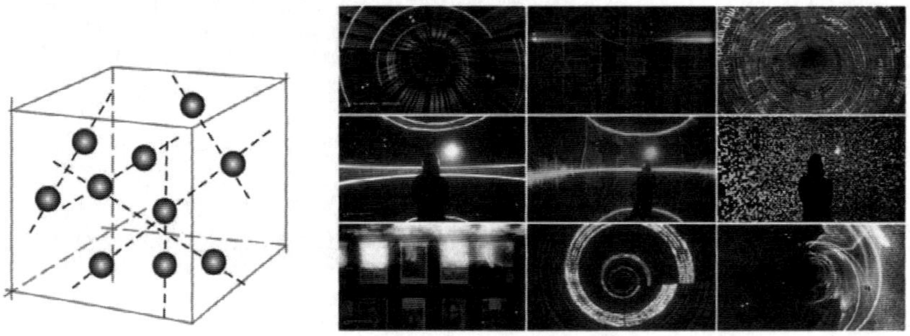

Figure 5. Model&Example.

SALT Galata Art Museum has lots of artworks. Different logic of design is needed to classify artworks, decide timeline and connection employed to present artworks, and change scenes. Different clues appears accordingly, that they affect each other to form a complicated net structure. Such kind of project always does not have a theme, and starts from dealing with lots of materials. But it will be finally linked together by many clues.

5. Relationship between immersion design and experience
Imagine development of relationship between users and an immersive environment as observation and learning done by human on their surroundings. It is similar to real space. During this process, it firstly relies on collection and analysis on visual information; take spatial awareness developed based on visual information for example, such cognitive model is also effective in virtual environment. Of course, construction of immersive environment is not limited to imitation on the reality. The nonlinear characteristics of digital media provide more possibilities: there is no physical rule in virtual reality, and form, location and scenography break through the limitation, which stimulate people's

imagination on spatial construction. Audiences will reflect unconsciously on the issue of "existence" when they are under such environment. What kind of role the individual play in such environment? Audiences' acceptance level is controlled by individual consciousness; information conveyed by vision will affect individual consciousness after being enhanced through form. During this process, level of influence is divided into several ones to do related presentation.

When we read original characters and pictures, we may be touched by contents, related images may be generated in mind, that we reach a concentrative status. This is a relatively low level of immersion, and its experience value is relatively low, too. Even the reading experience is well designed, its improvement is limited. We will pay more attention on dynamic frames when we watch TV; information transfer becomes rapider and more stereo. It is easy for people to accept no matter it is from the perspective of physiology or from the perspective of mentality; and people will not feel tired easily. Immersion reaches a higher level, reaching a dedicated status. However, such experience is related to quality of television programme, size of TV screen, and environmental comfort level. People will also get to this kind of status if they use computer to watch TV play or play games. Such immersion level will be improved further if people are in movie theater, because shadow casting technique, quality of sound source, and comfort level are greatly improved, people's experience will be improve greatly accordingly. This is why more people are willing to watch movie in movie theater.

The immersion mentioned throughout the paper refers that experience reaches the highest status, which is deep immersion. According to cases, to reach such status, technical support is absolutely necessary, that lamplight and projection realized with the help of computer are applied to immersive design frequently. Projections are either immersive, like a cinematic experience, or augmentative, in the sense of a mixed reality [11]. Lamplight could light up the darkness; and phototropism is animals' instinct. Light in darkness could make audiences have associations, and arouse people's memory in their heart. Different from common light source, lamplight used in immersive design is closely linked to computer, and becomes the method to reach high-class immersion status, which is arranged by designer in advance. Metaphors of vision and light are always combined with abstract concepts and immaterial concepts; however, in the era that it is filled with computer-generated imagery model, such two topics have been combined and have evolved to a relatively radical new dimensionality. Assisted by computer, projection and lamplight not only make people enter "deep immersion" status more easily, but also provide amicable user experience.

6. Conclusion

Computer, a kind of machine which seems to be widespread, is connected to human since it was invented. At the beginning, it was used to assist production. In 1980s, metaphorical interactive design of computer changed essentially, that modern graphical interfaces started to replace language codes. Interactive image design appeared. Immersion, a kind of marginal issue, attracts more and more attention. It is closely linked to visual effect and experience. Immersion is actually a kind of property of interactive image; designers amplify such property through some kind of method to enhance visual effect, information transfer and experience of design works.

References

[1] Slater M and Wilbur S 1997 *PRESENCE-TELEOP VIRT* **6** 603
[2] Riva G, Davide F and Ijsselsteign W 2003 *Being There: Concepts, Effects and Measurements Of User Presence in Synthetic Environments Studies in New technologies and Practices in Communication*(Amsterdam:IOS Press) pp 3-16
[3] Rettie R 2004 *7th Annual International Workshop on Presence* (Valencia) pp 117-124
[4] Kitson A, Prpa M, and Riecke B E 2018 *Front Psychol* **9** 1354
[5] Hansen M B N 2004 *New Philosophy for New Media* (Cambridge MA: The MIT Press) pp 8-11
[6] Manovich L 2001 *The Language of New Media* (Cambridge MA: The MIT Press) pp 108-112
[7] Grau O 2003 *Virtual Art: From Illusion to Immersion* (Cambridge MA: The MIT Press) p 13
[8] Xu J and Wu M C 2007 *Nonlinear Narrative* (Hangzhou: China Academy of Art Press)

[9] Meggs P B 1992 *Type and Image: The Language of Graphic Design* (Hoboken: John Wiley & Sons)
[10] Poulin R 2012 *The Language of Graphic Design: An Illustrated Handbook for Understanding Fundamental Design Principles* (Beverly MA: Rockport Publishers) pp 26-29
[11] Ernst S J G 2016 *The Magic of Projection: Augmentation and Immersion in Media Art* (Leiden: Leiden University Repository)

IOP Conf. Series: Materials Science and Engineering **520** (2019) 012017 doi:10.1088/1757-899X/520/1/012017

Parametric Design Method Based on Grasshopper and Shoe Last Bottom Pattern Moulding Characteristics

Yujing Tian[1], Yuxin Miao[1], Ying Yu[1,*] and Ziran Zhang[2]

[1] Donghua University, Fashion and Design College, 200051 Shanghai, P.R. China
[2] Shanghai University of Engineering Science, School of Art and Design, 201620 Shanghai, P.R. China

* yuyin513@163.com

Abstract. To adapt to the user's need for flexible and custom design of shoes, a more efficient customization method is explored. Combined with the formation characteristics of the shoe last bottom map, each characteristic parameter and related algorithm are analysed. Based on the importance of the shape control, the feature parameters are graded. The parametric design plug-in Grasshopper of Rhino software is used to construct the automatic moulding program of the shoe last bottom to form a professional and feasible design method. Through the time consumption comparison between the traditional method and the parametric design method, the efficiency advantage is analysed, so as to truly improve the design and production efficiency of the customized shoe last bottom.

1. Introduction

In the new age, people have had new and customized requirements for the shoes. How to design and process shoes quickly and flexibly has become a problem urgently to be explored by every shoe company[1]. Various custom design methods have also been developed that are used in all aspects of shoe design. The design of the shoe last bottom is an integral part of the design and production of any type of shoe, and its morphological features play an extremely important role in the aesthetics and comfort of the final shoe. In the shoe last bottom design process, quickly producing various types of shoe last bottom based on different needs has become the key to customization.

Parametric design is a non-linear design method allows you to quickly generate different design patterns by changing parameters in a visual programming language[2-3]. In the shoe last bottom design process, looking for characteristic parameters and morphogenesis logic, combined with parametric nonlinear thinking features, can effectively meet flexible design and production requirements[4]. Based on the Rhino parametric design plug-in Grasshopper, this paper analyses the forming characteristics of the last bottom, constructs the parameter logic, and forms a rapid prototyping method for the customized base.

2. Parameter analysis of traditional last bottom moulding characteristics

The shoe last bottom is based on the foot shape, and the shape of the shoe last bottom must follow the foot shape, so the design of the shoe last bottom is closely related to the parameters of the foot[5]. This article takes the length of female standard shoes as 230 mm as an example to briefly introduce the drawing steps of traditional shoe last bottom, as follows:

IOP Conf. Series: Materials Science and Engineering **520** (2019) 012017 doi:10.1088/1757-899X/520/1/012017

2.1. Calculation of shoe last bottom moulding characteristic point data

Based on the length of the foot, the reference value of the formed feature point of the last bottom was calculated from percentage relationship between foot length and parameters in national standards, it can be adjusted based on the measurement data of the foot type and design style in customization, including the length direction and the width direction, and a total of 16 related molded feature points were obtained (Table 1).

Table 1. Determination of last bottom moulding characteristic parameters of standard female shoe with a foot length of 230 mm.

Direction	Parameter	Calculation method	Length (unit: mm)
Length direction	Foot length	Measurement data (Known data)	230
	Posterior tolerance point	≈Foot length *2%	4.5
	Heel point	≈Foot length *18%	36.9
	Waist point	≈Foot length *41%	89.8
	The fifth metatarsal-phalange point	≈Foot length *63.5%	141.6
	The first metatarsal-phalange point	≈Foot length *72.5%	162.3
	Outer convex of little toe	≈Foot length *78%	174.9
	Outer convex of big toe	≈Foot length *90%	202.5
	Toe endpoint	≈Foot length – posterior tolerance	225.5
	Last toe allowance	Based on the design style, an intermediate value is given here	16.5
Width direction	Heel overall width	Based on the national standard reference data, it can be adjusted based on the measurement data of the foot type in customization.	50
	Waist outer width		32.5
	Fifth metatarsal-phalange outer width		43.9
	First metatarsal-phalange inner width		31.2
	Outer width of the little toe		43
	Inner width of the big toe		27.7
	Toe cap width	Based on the design style, an intermediate value is given here	30

2.2. Measurement of length characteristic markers

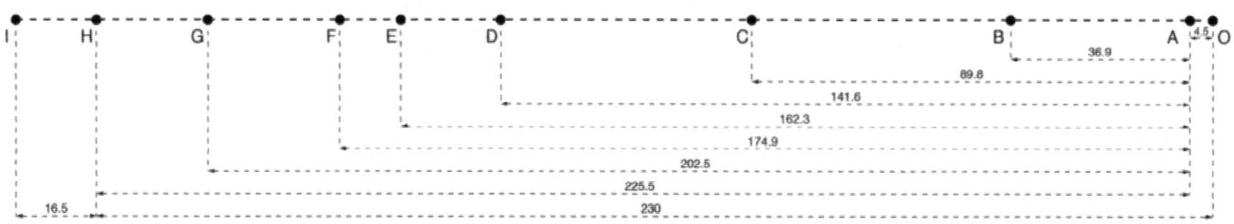

Figure 1. Measurement of length characteristic markers (unit: mm).

Draw a straight line as the bottom axis and measure the following dimensions on the shaft: OA, the length of posterior tolerance is 4.5 mm; AB, the length of heel point is 36.9 mm; AC, the length of waist is 89.8 mm; AD, the length of the fifth metatarsal-phalange point is 141.6 mm; AE, the length of the first metatarsal-phalange point is 162.3 mm; AF, the length of the outer convex of small toe is 174.9 mm; AG, the length of the outer convex of big toe is 202.5 mm; AH, the length of toe end is 225.5 mm; HI, last toe allowance is 16.5 mm, as shown in Figure 1.

2.3. Measurement of width characteristic markers

With the above point as the standard, the vertical line of the axis is as follows: the upper and lower parts of heel point are 25 mm, two points BB1, BB2, the full width of heel is 50 mm; CC', outer width of waist is 32.5 mm; DD', fifth metatarsal-phalange outer width is 43.9 mm; EE', first metatarsal-phalange outer width is 31.2 mm; FF', outer width of the little toe is 43 mm; GG', Inner width of the big toe is 27.7 mm, I1I2, toe cap width is 30 mm, as shown in Figure 2.

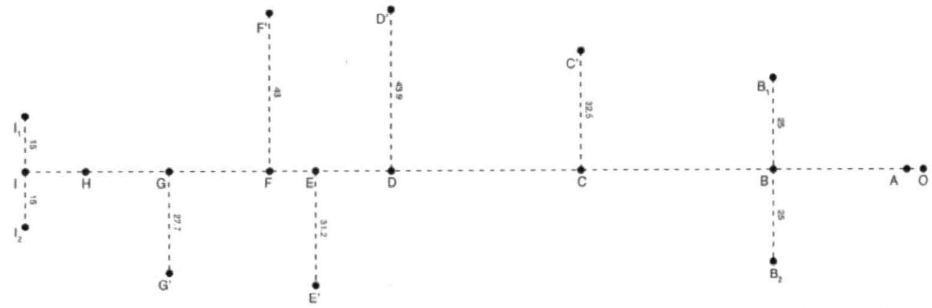

Figure 2. Measurement of width characteristic markers (unit: mm).

2.4. Determination of center line and heel width

The connection point of the branch line is determined by the equidistant relationship between the width of the first and the fifth metatarsal-phalange outer width, and D'J = EE' is measured on DD' to obtain the connection between points J and A. With A as the centre, the length of AB is the radius of the arc, AJ is taken as B', and B' is taken as the vertical line of AJ, and the two sides are divided into 25 mm, and B'B'1 and B'B '2 are obtained. B'1 B'2 is the width of heel, move B'1 B'2 to point A to get A1 A2, as shown in Figure 3.

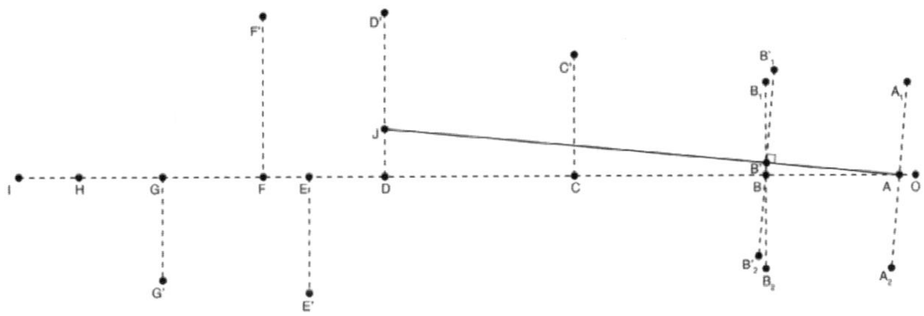

Figure 3. Determination of centre line and heel width (unit: mm).

2.5. Curve connection after straight line connection

As shown in Figure 4, these lines connect E, B'2, A2, A, A1, B'1, C', D', F', I, I1, I2, G', E' and C and the curve is connected to A, B'1, C', D', F', I, G', E' and B'2 based on the arrow, which is a female standard shoe map with a foot length of 230 mm.

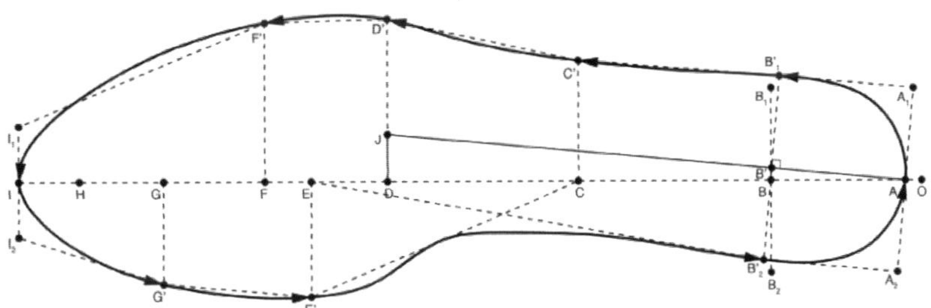

Figure 4. Curve connection after straight line connection.

3. Parametric logic construction based on Grasshopper

3.1. Hierarchical logic summary of characteristic parameters

Based on the importance level of the shape control of the parameter, the first level parameter, the second level parameter and the third level parameter are shown in Table 2. The first level parameter refers to the length of the foot, which is the basic parameter that should be determined first. The secondary parameter is a key feature parameter that is extracted through the standard drawing process, but it can actually be used to customize the shoe last bottom process. The measurements are modified; the third level parameters are based on the secondary parameters, and morphological control points are added for the morphological details to be used for the shape curve shape and the accuracy of the customization process. The corresponding step is to connect the key feature points to the last bottom curve in the shoe map. The shape of the connection curve is usually determined by the plotter's personal data experience and line aesthetics, so there is some uncertainty. The third-level parameters have no determined parameter values, and it is still necessary to adjust the curve by adjusting the points of each position with a certain experience. The logic organization is as shown in Figure 5.

Table 2. Parameter variables required for last bottom pattern moulding with Grasshopper.

Level	Parameter	Effect
Level 1	Foot length	Overall foot length controlled for different users
Level 2	Percentage of posterior tolerance point Percentage of heel point Heel point width Percentage of waist point Waist outer width Percentage of the fifth metatarsal-phalange point Fifth metatarsal-phalange outer width Percentage of the first metatarsal-phalange point First metatarsal-phalange inner width Percentage of outer convex of little toe Outer width of the little toe Percentage of outer convex of big toe Inner width of the big toe Allowance Toe cap width	Characteristic point of last bottom pattern moulding
Level 3	Heel control point	Control the shape of heel curve between posterior tolerance point and heel point
	Outer heel anterior shape control point	Control the shape of heel curve between outer heel point and outer waist
	Inner heel anterior shape control point	Control the shape of heel curve between inner heel point and first metatarsal-phalange point
	Fifth metatarsal-phalange posterior shape control point	Control the shape of heel curve between outer heel point and fifth metatarsal-phalange point
	First metatarsal-phalange posterior shape control point	Control the shape of heel curve between inner heel point and first metatarsal-phalange point
	Outer shoe shape control point	Control the shape of shoe curve between outer convex of little toe and toe cap
	Inner shoe shape control point	Control the shape of shoe curve between outer convex of big toe and toe cap

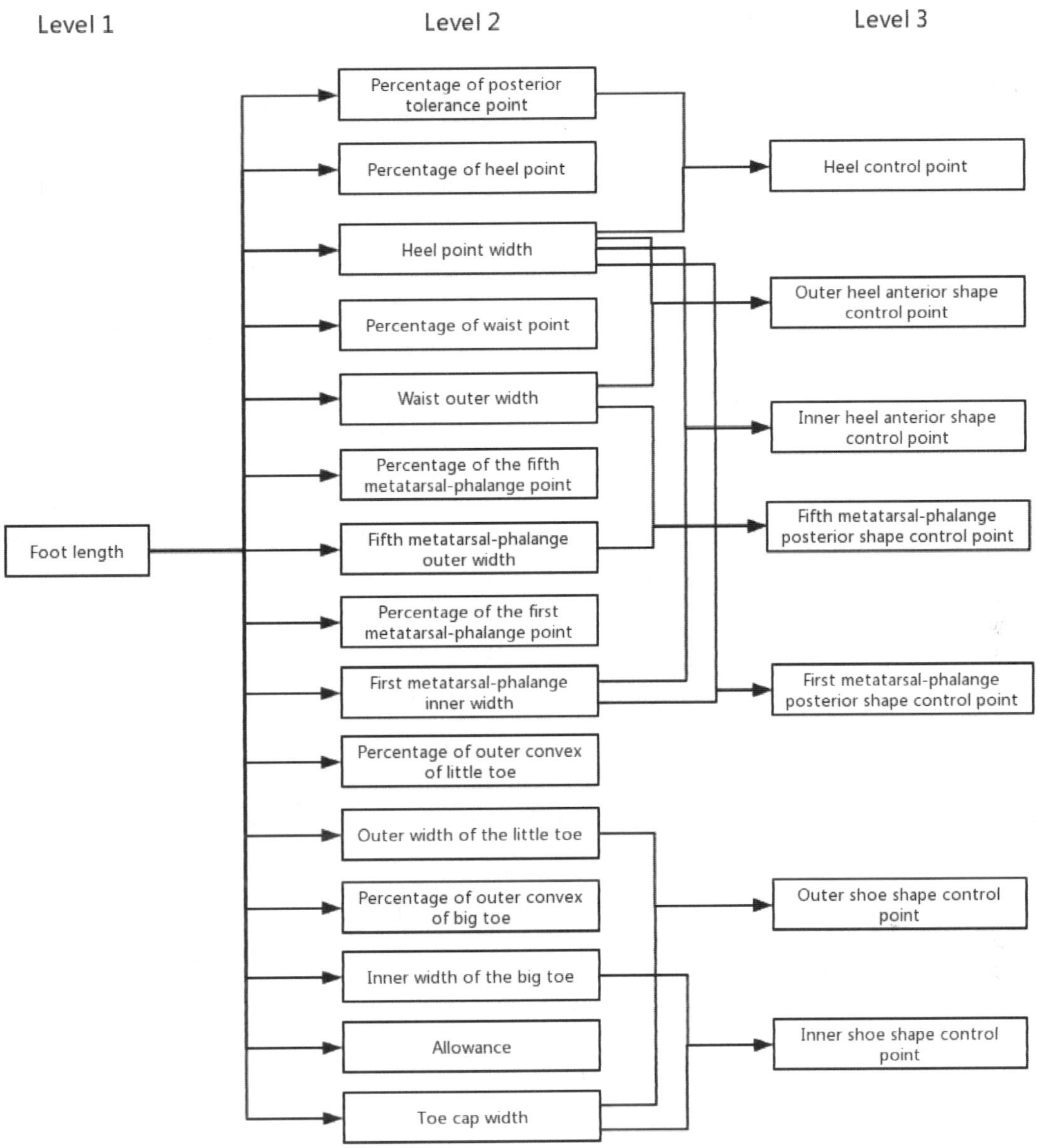

Figure 5. Level 1, 2 and 3 parameters.

3.2. Grasshopper edit input logic relationship

The above characteristic parameter parameters are logically coupled in combination with the moulding process of the conventional shoe last bottom pattern. It includes the length calculation method and shape connection method of each feature part. The bottom map is obtained based on the personal data of the custom measurement. The basic operation method is used for the logic construction in the three-parameter plug-in Grasshopper of the three-dimensional modelling software Rhino. During the construction process, you'll see the results immediately on the Rhino interface. The final logical relationship between the program interface and the bottom graph is shown in Figure 6.

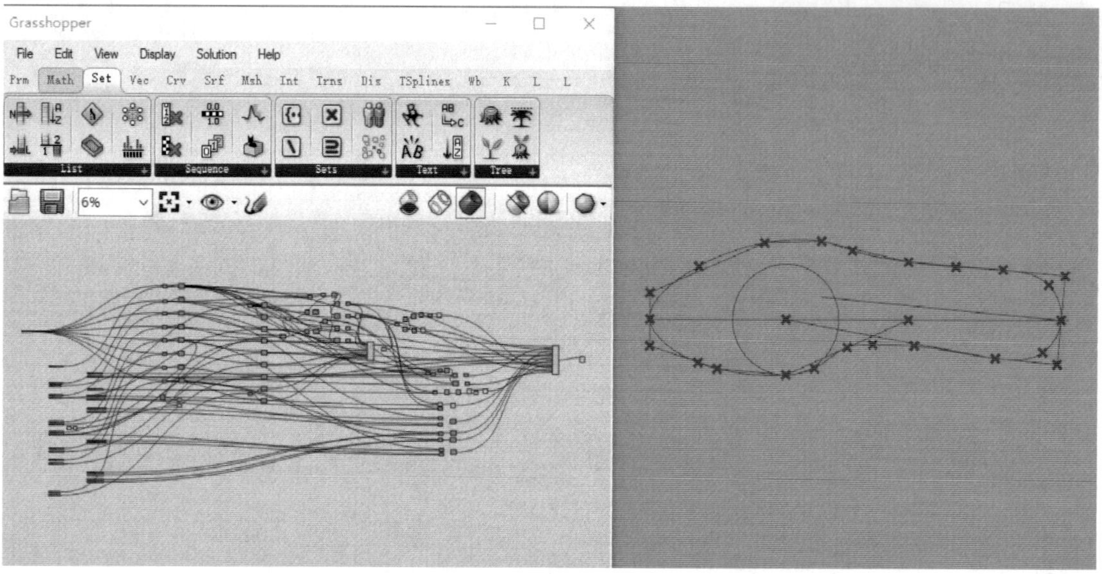

Figure 6. Grasshopper work interface and last bottom map.

3.3. Operation and inspection drawing effect

After modifying the shape of the curve by three levels of parameters to make the bottom map accurate, the final bottom curve will be baked separately to the Rhino interface, and you can open the edit point verification again, as shown in Figure 7. In order to facilitate the next 3D design and production of the last bottom, the corresponding format can be derived based on different moulding software.

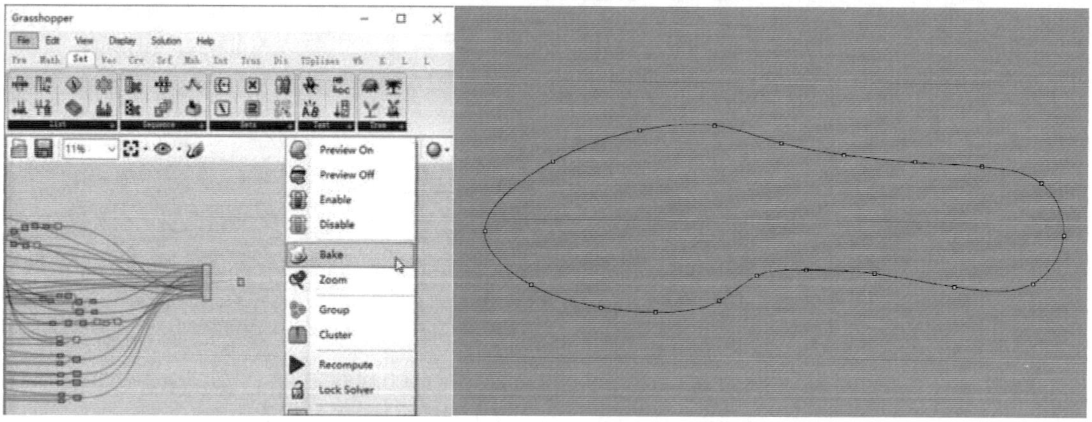

Figure 7. Bake the last bottom curve to Rhino in Grasshopper.

4. Comparative analysis of traditional and parametric last bottom drawing in shoe last bottom customization

4.1. Brief description of traditional last bottom drawing of shoe last bottom customization

Based on the traditional personal shoe last bottom process, the traditional drawing process steps are as follows:

(a) Measure the characteristic parameters of the target user's foot, including the foot length, percentage of heel point, heel point width, percentage of waist point, waist outer width, percentage of the fifth metatarsal-phalange point, fifth metatarsal-phalange outer width, percentage of the first metatarsal-phalange point, first metatarsal-phalange inner width, percentage of outer convex of little toe , outer width of the little toe, percentage of outer convex of big toe , inner width of the big toe.

(b) Determine the tolerance and width of the toe based on individual needs and styles.

(c) Based on the length of the foot, draw the bottom axis, and mark the position of the length direction of each point from the axis based on the measurement data.

(d) Mark the position in the width direction of each point based on the measurement data.

(e) Determine the allowance and shoe cap width.

(f) Connect each feature point with a polyline and calculate the remaining feature points based on the calculation.

(g) Connect each feature point with a smooth curve.

(h) Adjust the shape of the curve, curvature, etc. based on personal experience to make the shape more smooth and accurate.

(i) Scan the image and import it into a computer, then draw a curve for further 3D production.

Use the program you have written to complete the Grasshopper parameter shoe persistent drawing program. The steps to customize the personal basic map are as follows:

(a) Measurement parameters are the same as the conventional method.

(b) Determine the allowance and shoe cap width in the traditional method.

(c) The measurement data is directly input into the main and secondary parameters in the program, and the initial bottom curve is automatically generated.

(d) Adjust the position of the third-level parameters to make the curve shape more accurate.

(e) Direct output curve for further 3D production.

Compared to the parametric design process, the traditional drawing process requires a nine-step process, but the method of drawing the shoe bottom map is different. Due to the different experiences of the painter, in order to avoid the influence of the artist's experience, this article invites an experienced footwear primary student to use the above two methods for time-consuming comparative experiments. Finally, it was found that in the process of customizing the last bottom, the measurement of the characteristics of a single foot type is made to the production of an electronic version that can be further used. The basic map, the cumulative time of the traditional method is about 45 minutes; the parameterized bottom map method takes only 5 steps, about 11 minutes and 30 seconds, which greatly shortens the time of customizing the new bottom as shown in Table 3.

Table 3. Comparison of time consumption between traditional and parametric drawing methods.

	Time consumption	Traditional bottom drawing	Parametric bottom drawing
Measure the characteristic parameters of the target user's foot	5 min	√	√
Determine the allowance and shoe cap width	2 min	√	√
Based on the foot length, determine the bottom drawing axis, and mark the length direction of each point in the axis based on measured data	5 min	√	×
Mark the width direction of each point in the axis based on measured data	5 min	√	×
Determine the allowance and toe cap width	3 min	√	×
Connect each characteristic point with broken lines and obtain the remaining characteristic points based on the calculation	5 min	√	×
Directly input measurement data into the level 1 and 2 parameters in the program, automatically produce the preliminary bottom curve	2 min	×	√
Connect individual characteristic points with smooth curves	5 min	√	×
Adjust the curve shape, curvature based on	5 min	√	×

personal experience to make the shape smoother and more accurate			
Adjust the position of the level 3 parameters to make the curve shape more accurate	2 min	×	√
Scan and import image into the computer, and draw the curve for 3D production	10 min	√	×
Directly output curves for further 3D production	30 s	×	√
Cumulative time		45 min	11 min 30 s

5. Summary

Based on the moulding characteristics of the shoe last bottom and the parametric plug-in Grasshopper, the drawing method greatly improves the efficiency and ensures the accuracy of the form. In the future, individual foot parameters can be imported into actual production. The shoe last bottom is truly customized to ensure personal comfort. Such a combination is a technological innovation with commercial value, which will bring more possibilities for future shoe design.

Acknowledgments

Project Support by the Fundamental Research Funds for the Central Universities, Shanghai Style Fashion Design &Value Creation Collaborative Innovation Center & the Fundamental Research Funds for the Central Universities & Shanghai Summit Discipline in Design (DB18212).

References

[1] Pandremenos J, Georgoulias K, Chryssolouris G, Jufer N and Bathelt J 2016 a shoe design support module towards mass customization *Technology Management Conf.* (Trondheim: IEEE)
[2] Dai X 2016 the application and exploration of parametric design used in product design based on grasshopper *Design* 122-3
[3] Chen J 2015 *Art J.* 111-5
[4] He S and Liu B 2017 *J. Mech. Eng.* **53** 118-27
[5] Mahmoodi-Rad A, Molla-Alizadeh-Zavardehi S, Dehghan R, Sanei M and Niroomand S 2015 Int. J. Adv. Manuf. Technol. **85** 455-67

A Study on Unofficial Geographic Location Data Acquisition Technology Path under the Background of Big Data Era

Ziran Zhang[1, 2], Yujing Tian[3, a], Ying Yu[3] and Shengxi Fan[4]

[1] Tongji University, College of Architecture and Urban Planning, 200092 Shanghai, P.R. China
[2] Shanghai University of Engineering Science, School of Art and Design, 201620 Shanghai, P.R. China
[3] Donghua University, Fashion and Design College, 200051 Shanghai, P.R. China
[4] Tongji University, College of Design and Innovation, 200092 Shanghai, P.R. China

[a] Corresponding author: tinatian5186@163.com

Abstract. The arrival of the era of big data has made people break through the bottleneck of various technologies and resources faced by traditional social sciences, but at the same time, new problems such as data acquisition, data value mining, data storage, circulation and exchange, data privacy and security have also brought new challenges to people. In order to explore a new methodology of geographic location data acquisition, this paper proposed the technical path of unofficial geographic location data acquisition by using web crawler tools and test the feasibility of it.

1. Research background: the arrival of big data era

By 2020, the global data volume is expected to exceed 44 trillion gigabytes [1]. Such a huge amount of data also makes big data gradually become the focus of attention of all walks of life. The arrival of the era of big data has made people break through the bottleneck of various technologies and resources faced by traditional social sciences, but at the same time, new problems such as data acquisition, data value mining, data storage, circulation and exchange, data privacy and security have also brought new challenges to people.

1.1 The increasing concern of big data in China's research field

In 1980, Alvin Toffler, a well-known futurist in the United States, regarded data as the colorful movement of the third wave in his book The Third Wave. Although people did not realize how colorful it was at that time, it was regarded as the first reference to "big data". In 2001, Gartner Group analyst Douglas Laney first defined big data in terms of its characteristics and stressed the 3Vs of big data, namely volume, variety and velocity [2]. Viktor mayer-schonberger, a professor at Oxford University who is known as the father of big data in the industry, published the book Big Data: A Revolution That Transforms How We Work, Live, and Think in 2012 [3], and published a Chinese translation of the book in 2013, which initiated the research wave of big data in China. As can be seen from Figure 1, the development history of big data and the time that people pay attention to it are not long.

According to the statistical results provided by Baidu index, a popular search engine in China, its search volume in China gradually increased from 2013 to the highest around April 2017, and then entered a stable fluctuation state. Table 1 is the result of the author's literature statistics according to

IOP Conf. Series: Materials Science and Engineering **520** (2019) 012018 doi:10.1088/1757-899X/520/1/012018

the year after searching the keyword "big data" in China's well-known academic database "CNKI". The results show that after big data was clearly defined in 2001, the concept of big data did not attract extensive attention from the academic circle until 2011. In 2012, the research on big data has gradually become a research hotspot of Chinese scholars around such topics as big data era, big data technology, big data environment, big data analysis, big data industry and data mining. Its level of attention began in 2012, and increased sharply from 2013 to 2014, 4.7 times the number of documents, and 4.4 times in the five years from 2013 to 2018. Obviously, since 2013, big data has become a topic of widespread concern and the research object of scholars in China.

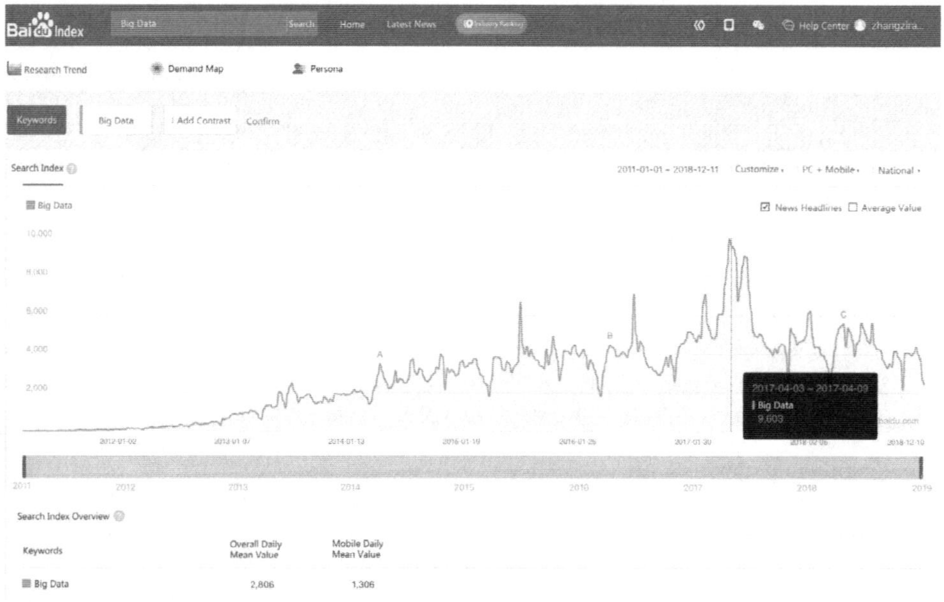

Figure 1. Hot trend of keyword "big data" search (Data source: Baidu index).

Table 1. Statistics on the number of Chinese papers on relevant keywords of "Big Data" on CNKI.

YEAR	BIG DATA	BIG DATA TIME	BIG DATA TECHNOLOGY	BIG DATA EVIRONMENT	BIG DATA ANALYSIS	BIG DATA INDUSTRY
2009	55	0	0	0	0	0
2010	69	1	0	2	0	0
2011	127	8	0	1	8	0
2012	994	214	47	24	55	20
2013	4739	1245	261	173	239	92
2014	10474	2539	600	590	414	312
2015	16708	3415	1020	1224	661	521
2016	22511	4029	1239	1925	876	634
2017	21070	4257	1735	2333	1030	666
2018	25774	4646	1909	2246	1014	468

1.2 Research status and development trend of location big data in China

As the main technology of mobile location data acquisition, LBS has reached a peak between 2012 and 2014 in China with 314 relevant Chinese academic papers in 2014. After 2013, it began to decline gradually, and it was sharply reduced to 45.2% of its peak from 2016 to 2018. This may be due to the gradual shift of people's attention from previous technology research and development to other directions because of the gradual maturity of LBS technology. At the same time, the research on location data has begun. It is not difficult to see from Figure 2 that compared with the two keywords "location data" and "geographical location information", the number of documents containing "big location data" and "mobile signaling data" in the title is relatively small. It started around 2014 and it is now in a slow climbing stage. However, compared with the three related documents in 2013, "big location data" has produced about 10-fold growth in quantity in 2018, and its development momentum cannot be underestimated.

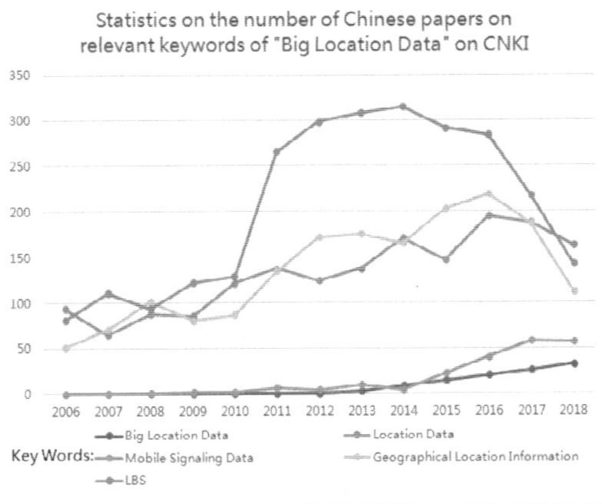

Figure 2. Statistics on the number of Chinese papers on relevant keywords of "Big Data" on CNKI.

1.3 Relations between geographical location data and location big data

Location data, as the key data to show people's moving trajectory, can bring great research value and business opportunities for transportation, tourism, population, urban development and other research directions. The relationship between geographic location data and location big data is a deeper rational discussion of urban population flow, people's spatial behavior, spatial structure and preferences and other issues and phenomena. For example, Tencent used Tencent big data and park location data to conduct a series of analyses and research on Shanghai park tourism preference during the National Day holiday in 2018, which provided strong support for young people's vacation tourism preference research. Liang Lin and other scholars established a network model for the characteristics of inter-city population flow in Beijing, Tianjin and Hebei based on Tencent location data [4]; Bai Yan, Zhu Anran and others explored the spatial distribution characteristics of night life in Hefei city based on the location data of Weibo[5]. Thus, it can be seen that selecting appropriate geographic location data according to research needs and combining location big data for relevant analysis and research have certain practical value and research significance for urban spatial structure planning and development.

2. Acquisition methods of geographic location data

If you want to complete a data project, data preparation often takes up 70% or more of the workload.[1] It is also known as "data cleaning" in the industry, *which is one of the 19 research topics on data quality* [6]. Usually, *instance-level data quality problems include similar duplicate records* [7], incomplete records, logical errors, abnormal data, etc. General speaking, *data cleaning is usually an expensive process that requires significant resources, considerable effort, and human interaction* [8]. Therefore, how to obtain relatively "clean" and valuable geographic location data to reduce the cost of cleaning data depends on the choice of data sources and data acquisition technology path.

2.1 Data acquisition methods in traditional social sciences

In the era of underdeveloped information technology, researchers of social sciences usually use computer-aided manual statistics to collect data. Geographical location and its related attribute information are very important research objects in the fields of landscape design, environmental art design and geography, sociology, tourism, traffic information technology, etc. Due to the immaturity of electronic map, remote sensing satellite technology and GIS algorithm, manual statistics or GPS positioning assisted manual statistics combined with scientific sampling method were used to obtain the specific location of a certain type of building and related attribute information for scientific research.

2.2 Platform paid data purchase

In recent years, due to the rapid development of big data technology in China, many commercial organizations have begun to gradually promote a series of data purchase services. For example, "Tencent Location Big Data" can directly sort out the location data of Tencent users and related attribute data according to the needs of buyers, and then trade or conduct more in-depth business cooperation. If you search Baidu for the keyword "Community Location Data Download", you can find many suppliers of data purchase, including tdata.cn, Shenjian.io, JDCloud, CSDN, etc. Through sample analysis and survey, the average purchase price of community location data in Shanghai is 2,000 Yuan, and the data purchased cannot guarantee its validity, credibility and the last update time, so the data quality is difficult to control.

2.3 Network information data source and crawler technology

The rapid rise of China's Internet economy in the past decade has promoted the maturity and perfection of network information technology in the domestic market of traditional industries. As third-party online platforms such as Dianping, Meituan, Taobao and Jingdong have changed from "being generally accepted by consumers" to "changing consumers' living habits", more and more merchants are forced to choose these third-party online platforms as the channels of customer flow. Therefore, the unshakable commercial status of these platforms is created, and all the information and massive data of consumers and businesses are centralized. Some of these data are published on such platforms' websites to facilitate consumers to make commodity selection decisions as very valuable data resources.

Crawler technology, in simple words, is a technology that can effectively extract web page data according to the logic specified by the program design. With the crawler technology and relatively centralized data resources, China's Internet environment has provided researchers with very favorable research conditions. Nowadays, this way of extracting and analyzing network data by using crawlers has been widely applied in the fields of network consumption preference, hot issues public opinion research and logistics industry.

3. Exploration of unofficial geographical location data acquisition technology path

3.1 Principles for selecting data sources

In order to obtain complete and reliable data, the source of data is particularly important. Usually, according to the research needs, 2-3 websites containing the information needed by the research are selected to extract the crawler data. The website needs to be widely accepted and used, and the amount of data must be nearly complete, or it can be selected and decided according to the market share or the number of users of the website. This is the big data mentioned above. After collecting and cleaning the data in the selected network platform, they can complement and merge with each other, and compare the data results with the data on other websites to test their reliability and validity.

3.2 Technological feasibility exploration: a case study of related attribute information collection of community in Shanghai

3.2.1 Selection and comparison of geographic location data sources in communities. In China, real estate agencies, as the private enterprises that know the situation of the surrounding communities best, have a large amount of specific information about the communities around the physical store outlets, including the address, construction date, number of residents, housing structure and selling price, etc. Therefore, if the official websites of real estate agencies with high coverage of online stores and large number of users are chosen as data sources to collect location and related information of residential areas, their data integrity and reliability are better. Major housing intermediary brands have their own free official websites in China, and their information content is updated in real time. Therefore, in this case, keywords "Shanghai Second-hand Housing" were searched on Baidu, and the data of the top 8 real estate agencies' official websites in the first 3 pages of the search results, except the advertising links, were taken as the data source for comparison. The statistical results are shown in Table 2 below. The top three with the largest amount of data were selected as data collection objects and the data information of all the search results in each administrative region under the "community" section was collected. Among them, the website information with the largest amount of data is regarded as the website with the most complete information, and it can be used as the main data source for data acquisition. In the statistical process, missing data information or information without search results was replaced by "0". In addition, Shanghai merged Nanhui District into Pudong New District in 2009. In 2011, "Luwan District" and "Huangpu District" were merged, and in 2018, "Zhabei District" and "Jing'an District" were merged. Therefore, relevant information was also merged in this statistics.

In Table 2, according to the search results, 58.com collected the most information items in Baoshan District, Songjiang District, Hongkou District, Qingpu District and Jinshan District, and they were highlighted in Table 2. Similarly, Fang.com searched for the largest amount of information in six administrative districts, including Pudong New District, Minhang District and Putuo District, and they were also highlighted. However, the highlighted data in the column of Fangdd.com were also the four regions with the largest number of search results compared with the corresponding administrative divisions of other websites. Therefore, this study selected 58.com, Fang.com, Fangdd.com as data sources to collect information on the geographical location of all residential districts in Shanghai, and added 660 district information collected by the "Fengxian District" on Lianjia to the information collected by Fengxian District on 58.com, in order to ensure the integrity of the data as far as possible.

Table 2. Community information statistics of Baidu's top 8 real estate intermediary websites in districts.

District	58.com	Fang.com	Lianjia	Anjuke	qfang.com	5i5j.com	CENTALINE PROPERTY	Fangdd .com
Pudong	2214	3869	3364	3285	1013	3201	2662	3500
Minhang	1655	1739	1596	1480	113	1430	796	1581
Xuhui	1675	1743	1592	1509	6	1455	917	1817
Putuo	1016	1047	985	936	429	960	415	981
Baoshan	1153	1034	1056	1052	415	793	823	1134
Changning	1213	1388	1252	1074	62	1150	745	1236
Yangpu	1290	1205	1342	1174	22	1047	878	1425
Songjiang	1128	898	963	998	0	696	212	995
Hongkou	1315	1150	1185	1076	14	942	849	1315
Jiading	1091	922	1074	999	35	768	550	1141
Huangpu	1397	1227	1321	1143	16	1054	659	1512
Jingan	1777	1883	1653	1547	435	1490	806	1809
Qingpu	806	566	632	752	0	485	91	707
Fengxian	640	477	660	577	0	0	145	446
Jinshan	426	358	330	379	0	0	133	307
Chong Ming	214	222	170	203	0	0	2	145
SUM	19010	19728	19175	18184	2560	15471	10683	20051

3.2.2 The use of web crawler tools and the setting of grabbing technical routes. In this paper, the working principle of crawler software has been introduced in 2.3. This study is to select one of the mature and free crawler software GooSeeker to download website data in batches. The main grab technology roadmap is shown in Figure 3. The main key steps are as follows: 1. Open the designated website; 2. Set cyclic select the designated administrative region; 3. Set cyclic page turning; 4. Grab the attribute information such as "community name", "address" and "building age" in the page (the specific attribute information can be set according to the research purpose). Among them, 51.com downloaded17781data, Fang.com and Fangdd downloaded 11861and 18502data respectively, and the statistical results are shown in the column of "actual downloads" in Table 3.

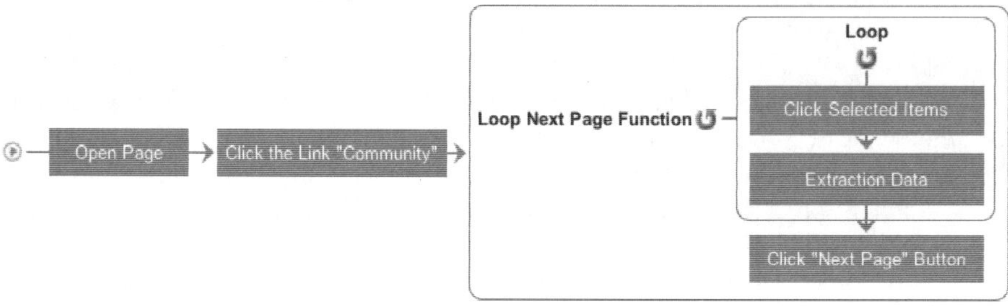

Figure 3. Logical schematic diagram of GooSeeker's main grabbing technology route.

Table 3. Statistics of data downloads from various real estate websites.

District	58.com			Fang.com			Fangdd.com		
	Planned Download	Actual Download	Data Collection Efficiency	Planned Download	Actual Download	Data Collection Efficiency	Planned Download	Actual Download	Data Collection Efficiency
Baoshan	1153	1090	94.54%	1034	707	68.38%	1134	1134	100.00%
Chongming	214	208	97.20%	222	187	84.23%	145	145	100.00%
Fengxian	640	640	100.00%	477	333	69.81%	446	444	99.55%
Hongkou	1315	1315	100.00%	1150	773	67.22%	1315	1309	99.54%
Huangpu	1397	567	40.59%	1227	667	54.36%	1512	1506	99.60%
Jiading	1091	1061	97.25%	922	528	57.27%	1141	1137	99.65%
Jinshan	426	426	100.00%	358	250	69.83%	307	306	99.67%
Jingan	1777	1750	98.48%	1883	1136	60.33%	1809	1799	99.45%
Minhang	1655	1570	94.86%	1739	1064	61.18%	1581	1574	99.56%
Pudong	2214	2104	95.03%	3869	2000	51.69%	3500	1997	57.06%
Putuo	1016	980	96.46%	1047	644	61.51%	981	980	99.90%
Qingpu	806	802	99.50%	566	295	52.12%	707	705	99.72%
Songjiang	2206	1128	51.13%	898	411	45.77%	995	988	99.30%
Xuhui	1675	1674	99.94%	1743	1094	62.77%	1817	1823	100.33%
Yangpu	1290	1253	97.13%	1205	893	74.11%	1425	1420	99.65%
Changning	1213	1213	100.00%	1388	879	63.33%	1236	1235	99.92%
SUM	20088	17781	88.52%	19728	11861	60.12%	20051	18502	92.27%
AVERAGE	1255.5	1111.3125	91.38%	1233	741.3125	62.74%	1253.1875	1156.375	97.06%

3.2.3 Data cleaning and verification. When GooSeeker grabs data, it stores each page data as XML format file separately. Therefore, the number of XML format files in the data storage folder is the number of pages successfully grabbed. Statistical comparison shows that under the premise of accurate field design, the page crawl rate is 91.38%, 62.74% and 97.06%, respectively. As shown in Table 3, Page crawl rate is divided actual crawl data to planned crawl data. According to the commonly used data cleaning methods, the following steps of data cleaning are carried out by software: 1) convert XLM files into CSV data format files; 2) delete duplicate and invalid data, including invalid data items of "shops, office buildings, etc;" 3) convert address information in data into longitude and latitude data through Baidu Map API access; 4) count the amount of data downloaded by various administrative districts and compare it with the planned amount of data downloaded to ensure that there is no missing data.

IOP Conf. Series: Materials Science and Engineering **520** (2019) 012018 doi:10.1088/1757-899X/520/1/012018

3.2.4 Verification of data. The verification of the geographical location information data of the community can be transformed into two questions, namely, whether the "community name" in the data collected by each website is uniform and whether the address information of the community is correct. If the consistency is high and the correct rate of address information is high, the data acquisition method can be regarded as effective. Otherwise, it is not.

E15 =ISNUMBER(MATCH(C15,A:A))

	A	B	C	D	E	F	G
1	Name of Community From Different Real Estate Agents			Consistency			
2	Fangdd.com	58.com	Fang.com	A-B Consistency	A-C Consistency	B-A Consistency	C-A Consistency
3	Songnan Fifth village	Central No.1(First Item)	Central No.1	TRUE	TRUE	TRUE	TRUE
4	Tianxin Garden	Siji Luchen(South District	Songnan Fifth Villag	TRUE	TRUE	TRUE	TRUE
5	Dahua First village	Sanhua Xiandai City Jinlan	Ploy Ye Shanghai(Apa	TRUE	TRUE	TRUE	TRUE
6	Dahua Second village	Shenshi Baodi(First Item)	Tonghe First Village	TRUE	TRUE	TRUE	TRUE
7	Haishang Mingcheng	Baohong Jiayuan	Hexin Garden	FALSE	FALSE	TRUE	TRUE
8	Vanke Wonderland	Meian Xiting	Taihe Xin Chen	TRUE	TRUE	TRUE	TRUE
9	Meian Xiting(Apartment)	Tianxin Garden(Apartment)	Gongkang Fifth Villa	TRUE	TRUE	TRUE	TRUE
10	Taihe New City	Jingqiu Garden(Apartment)	Dahua First Village	TRUE	TRUE	TRUE	TRUE
11	Yuanyang Xiangnai	Wenbao Yuan	Tonghe Second Villag	TRUE	TRUE	TRUE	TRUE
12	Meiluo Jiayuan	Songnan Fifth Village	Baochen Gonghe Jiayu	TRUE	TRUE	TRUE	TRUE
13	Jiajing yuan	Hexin Garden	Baochen Yijing Yuan	TRUE	TRUE	TRUE	TRUE
14	Tonghe Second Village	Baochen Yijing Yuan	Luonan Second Villag	FALSE	FALSE	TRUE	TRUE
15	Xinjia Yuan Sixth Block	Dahua First Village	Huma Third Village	TRUE	TRUE	TRUE	TRUE
16	Songnan Seventh village	Poly Leaf(Apartment)	Jinqiu Huayuan	TRUE	TRUE	TRUE	TRUE
17	Longhu Beicheng Tianjie	Baoshan Wenhua Yuan(Apartm	Dahua Shuian Lanqiao	TRUE	FALSE	TRUE	TRUE
18	Zhongyi Jiayuan	Huma Thrid Village	Ludi Gongyuan Yipin	TRUE	TRUE	TRUE	TRUE
19	Siji Luchen	Binjiang Yayuan(First Item	Zhongguo Tiejian Qin	TRUE	TRUE	TRUE	TRUE
20	Ploy Ye Shanghai(Apartment)	Bojue Gongguan	Meilan Hupan lanya Yu	TRUE	TRUE	TRUE	TRUE
21	Gubei Juxiang Yuan	Dahua Shuian Lanqiao(Apart	Fengshui Baodi Xiyua	TRUE	TRUE	TRUE	TRUE
22	Gaojing First Village	Jufeng Jingdu	Jindi Tiandi Yunshu	TRUE	TRUE	TRUE	TRUE
23	Rd Jipu 615	Gongfu Second Village	Meilan Hu Yijing Yua	TRUE	TRUE	TRUE	TRUE
24	Wenbao Yuan	Meilan Hu Yijing Yuan	Gongkang Seventh Vil	TRUE	TRUE	TRUE	TRUE
25	Yangtai Second Village	Dahua Bojin Huafu	Tianxin Hua Yuan	TRUE	TRUE	TRUE	TRUE
26	Jinggao Second Village	Gongfu First Village	Gongkang Sixth Villa	TRUE	TRUE	TRUE	TRUE
27	Xinjia Yuan eleventh Block	Jingwei Xuefu Hanqin	Luzhou Garden	TRUE	TRUE	TRUE	TRUE
28	Sitang Fifth Village	Caiju Yuan	Meilan Hu Zhonghua Yu	TRUE	TRUE	TRUE	TRUE
29	Huma Third Village	Meilan Hu Zhonghua Yuan	Wanke Hupo	TRUE	TRUE	TRUE	TRUE
30	Huabing New Village	Xuelin Yuan	Jingwei City Oasis(S	TRUE	TRUE	TRUE	TRUE
31	Gongfu First Village	Baoshan Third Village	Tonghe Eighth Villag	TRUE	FALSE	TRUE	TRUE
32	Juquan New City(Rd. Luxiang No.3	Ploy Ye Shanghai(Apartment	Dongfang Pati Ou Cit	TRUE	TRUE	TRUE	TRUE
33	Juquan New City(Rd. Jushen No.50	Gaojing First Village	Zhonghuan Guoji Apar	TRUE	TRUE	TRUE	TRUE
34	Kangqiao Shuidu(Apartment)	Meilan Hu Lingyu(Apartment	Jingwei City Oasis(F	TRUE	TRUE	TRUE	TRUE
35	Xinjia Yuan twelfth Block	Baotong Jiayuan	Tonghe Sixth Village	TRUE	FALSE	TRUE	TRUE
36	Wanye Zichen Yuan	Haijiang Second Village	Dahua Second Village	TRUE	TRUE	TRUE	TRUE
37	Tonghe Sixth Village	Xuelin Yuan	Jingrui Lixiang Zhigu	TRUE	TRUE	TRUE	TRUE
38	Jingwei City Oasis(First Item)	Xin Meisong Nanyuan	Siji Luchen	TRUE	TRUE	TRUE	TRUE
39	Tonghe First Village	Hao Rizi Dajia Yuan Distri	Poly Yeyu (First Ite	TRUE	TRUE	TRUE	TRUE
40	Qianxi New Village	Baogang Baolin First Villa	Dahua Bojin Huafu	FALSE	FALSE	TRUE	TRUE
41	Qilian Second Village	Xuhui Lanyue Wan	Yuanyang Xiangnai	TRUE	TRUE	TRUE	TRUE
42	Huma Third Village	Luzhou Garden	Gaojing Second Villa	TRUE	TRUE	TRUE	TRUE
43	Shenshi Baodi(First Item)	Gongkang Fifth Village	Kangtai New City	TRUE	TRUE	TRUE	TRUE
44	Huma Second Village	Baoyue Jiayuan	Huma Second Village	TRUE	FALSE	TRUE	TRUE

Baoshan / Chongming / Fengxian / Hongkou / Huangpu / Jinshan / Jingan / Minhang / Putuo / Qingpu / Songjiang / Yangpu / Changning

Figure 4. Comparisons of community data consistency (Baoshan District).

Fig. 4 is an illustration of the sample data used for inspection after cleaning, sorting and merging. Column A is the data column to be tested, and the data source of column A is formed by combining the data with the largest amount of downloaded data in the corresponding administrative region of each website, namely the data marked as high-light color in Table 3 (For example, the community location information data of Baoshan District, Chongming District and other administrative regions downloaded from 58.com; Data of Huangpu District, Jiading District, etc. downloaded from Fangdd). The community name is extracted and merged into the column A data here, and the consistency test is conducted for this column data and other data according to the administrative region. If column A of the data to be tested within the designated administrative region can largely cover the data downloaded from the other two websites (data from the control group), the information integrity and credibility of column A of the data to be tested are relatively high.

Table 4 is the consistency test results of community names in the data of each website. The consistency is divide the total number of communities with the same community name in column B or C as in column A to the total number of communities in column B or C. From the results, it is not difficult to find that the consistency between the tested data columns (column A) and the experimental group data columns (B, C columns) is high. The average consistency is between 95.36% and 96.53%, and the highest consistency is 100% in Baoshan, Chongming and other districts. That is to say, column A contains all the data in column B and column C. In this test, the lowest consistency occurs in Jiading and Jinshan District, and column A contains 84.83% and 85.98% data in column B and column C respectively.

IOP Conf. Series: Materials Science and Engineering **520** (2019) 012018 doi:10.1088/1757-899X/520/1/012018

Table 4. Consistency test results of community names in the data of websites.

District	B-A Consistency	C-A Consistency	A-B Consistency	A-C Consistency
Baoshan	99.91%	100.00%	94.11%	91.68%
Chongming	100.00%	100.00%	86.45%	97.20%
Fengxian	100.00%	100.00%	89.84%	95.00%
Hongkou	100.00%	100.00%	88.14%	83.73%
Huangpu	88.71%	87.56%	100.00%	89.31%
Jiading	84.83%	85.98%	100.00%	96.14%
Jinshan	84.83%	85.98%	100.00%	96.14%
Jingan	86.82%	86.09%	100.00%	98.45%
Minhang	100.00%	100.00%	96.98%	97.28%
Pudong	100.00%	100.00%	95.53%	85.36%
Putuo	100.00%	100.00%	99.70%	92.42%
Qingpu	100.00%	100.00%	95.29%	95.29%
Songjiang	100.00%	100.00%	98.37%	99.27%
Xuhui	99.64%	99.73%	100.00%	98.41%
Yangpu	91.01%	89.81%	100.00%	95.08%
Changning	89.95%	89.42%	100.00%	93.68%
Average	95.36%	95.29%	96.53%	94.03%
Variance	0.003738991	0.003758881	0.002019555	0.001944446

4. Conclusions

4.1 Technical path of unofficial geographic location data acquisition

To sum up, with regard to the collection of unofficial geographic location data, it is necessary to classify and download the data on the basis of selecting the right data source, and select multiple data download channels at the same time to use crawler technology to download and form data sample 1, data sample 2 and data sample 3…, combine the samples with the largest amount of data in the data sample to form the data to be tested, take the remaining sample data as the data of control group, use the API import function of the map software to convert the address information in the complete data that passes the consistency test into the longitude and latitude information for researchers to use. The flow chart can be summarized as Figure 5.

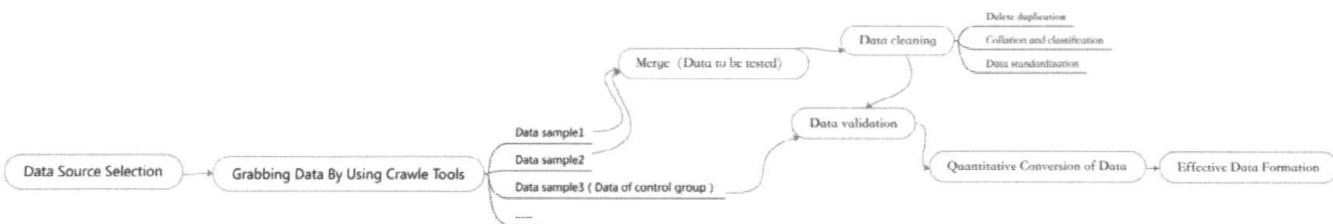

Figure 5. Technical path of unofficial geographic location data acquisition.

4.2 Comparison of several geographical location data acquisition methods

Table 5 shows the differences of economic cost, time cost, technical difficulty and feasibility of several different geographic location data acquisition methods, which are expressed in three grades: high, medium and low. Among them, the traditional manual acquisition method has the highest cost, but because it has no high requirements for technology, it is still used today. Compared with time-saving data purchase service, it also has low technical requirements. However, because data service is a commodity publicly sold by network technology enterprises nowadays, its economic cost is relatively high, it is impossible to grasp the reliability and integrity of data by oneself. Finally, although the method of collecting and processing unofficial data sources by using crawler technology has certain requirements for technology and requires researchers to learn crawler software by themselves, most crawler software on the market have preset templates for users to choose, so they can quickly learn to operate without programming. Therefore, this method provides a new technical path for researchers who are short of money and time for research.

Table 5. Comparison of several different geographic location data acquisition methods.

	Economic Cost	Time Cost	Technical Difficulty	Feasibility
Data acquisition methods in traditional social sciences	High	High	Low	Medium
Platform paid data purchase	High	Low	Low	Unknow
Network information data source grab with crawler technology	Low	Low	Medium	High

Acknowledgments
Project Research Team of Intelligence Sustainable Package Design Support by Shanghai Summit Discipline in Design; Foundation item: Chenguang Plan (13CG74) sponsor; Shanghai Style Fashion Design &Value Creation Collaborative Innovation Center Support by Shanghai Summit Discipline in Design (DB18212)

References

[1] KPMG China big data team 2018 *From Data to Insights* (Beijing: Tsinghua University Press) p 01
[2] Jing L 2018 The origin and action of big data, *Economics Think Tank* **04** 26-35
[3] Mayer-Schönberger V, Cukier K 2013 *Big Data: A Revolution That Transforms How We Work, Live, and Think* (Boston: Houghton Mifflin Harcourt)
[4] Liang L, Zhao Y B and Liu B 2019 *Northwest Population Journal* **40** 20-28
[5] Bai Y, Zhu A R, Yang Y Y, et al 2018 Spatial Study of Nightlife in Hefei Urban Area Based on the Place Data in Micro-blog *Architecture and Culture* **175** 185-186
[6] Madnick S E, Wang R Y 2009 Overview and Framework for Data and Information Quality Research *ACM Journal of Data and Information Quality* **1** 1-22
[7] Cao J J, Diao X C, Tan M C, et al, 2010 *the 15th International Conference on Inforamtion Quality* (Arkansas USA: UALR)
[8] Cao J J, Diao X C 2017 *Introduction to Data Quality* (Beijing: National Defense Industry Press) p 42

IOP Conf. Series: Materials Science and Engineering **520** (2019) 012019 doi:10.1088/1757-899X/520/1/012019

A Novel Clothing Design Method for Children with Autism Spectrum Disorders Based on Fun Computing Theory and Substance-field Analysis

Xin Chen[1] **and Tao Shen** [2, a]

[1] School of Fashion, Dalian Polytechnic University, Dalian, China
[2] Knowledge Science, Japan Advanced Institute of Science and Technology, Nomi City, Japan

[a] Corresponding author: shentao@jaist.ac.jp

Abstract. Autism spectrum disorders (ASD) is a natural disorder that is characterized by impaired communication, impaired interaction and also by limited and repetitive behaviour. In this paper, we present a novel method based on Fun Computing and Substance-Field (Su-field) Analysis to empower designers in conceiving effective clothing design making children with ASD more interactive. In this method, Fun Computing is used to enhance the children's enthusiasm for interaction and Su-field analysis contributes to describing the interactions and behaviours in the design process. Finally, we conduct a controlled experiment to demonstrates the effectiveness of this method by measuring idea quality and quantity.

1. Introduction

Autism spectrum disorders (ASD) is a subtype of extensive developmental disorder, which is more common in men and occurs in infants and young children [1]. It is mainly characterized by varying degrees of speech developmental disorders, interpersonal barriers, narrow interest and behavioural stereotypes. About three-quarters of patients have significant mental retardation, and some have better abilities in some aspects of general mental retardation [2]. Individuals with ASD suffer direct and indirect consequences related to social interaction deficits. Children with ASD often have a desire for more peer social interaction, and may also express poor social support and more loneliness than their typically developing peers [3]. To improve this situation, interventions based on principles of applied behaviour analysis (APA) have been presented to improve functional communication skills in children with ASD and decrease problematic behaviours such as aggression [4][5], but healing interventions targeting social deficits have not achieved the same level of attention [6]. Indeed, social deficits in this population remain a main treatment challenge [7].

Meanwhile, design activities are the refinement, processing and re-creation of society. Designers extract valuable and meaningful contents from society and apply them to design activities to enrich people's lives and bring innovation to people. However, little attention is focused on the social care for handicap children such as children with ASD in clothing design, most clothing designers pay much attention to fashion, smart technology, marketing and so on. Herein, a novel method based on Fun Computing and Su-field Analysis are presented to empower designers in conceiving effective clothing design making children with ASD more interactive.

Our method is based on the understanding of fun computing and Su-field Analysis, Su-field Analysis is a TRIZ analytical tool for modelling problems related to existing technological systems [8]. Thus, this method is also considered as a tool which abbreviates the knowledge distance between designers' own knowledge and new ones in the total technology space [9].

2. The novel clothing design method for children with ASD

In TRIZ theory, Su-field Analysis is an analytical tool for modelling problems related to existing technological systems. The desired function is the output from an object or substance (S1), caused by another object (S2) with the help of some means (types of energy, F) [9,10,11]. Figure 1 shows the basic model of Su-field Analysis.

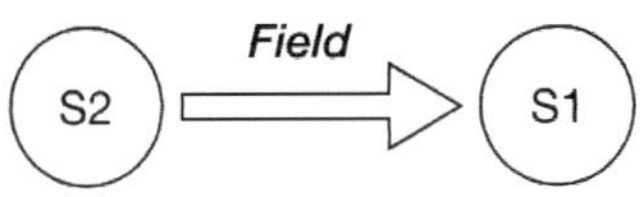

Figure 1. The basic model of Su-field Analysis

2.1. Fun computing theory

For clothing design, we define field is fun computing. 'Fun' is the easiest way to change people's behaviour for the better [12]. For example, some researchers installed giant piano keyboards covering the stairs leading out of a subway station in Stockholm. When people step on it, these keyboards play beautiful melodies. As a result, 66% more people than normal chose the stairs over the escalator, that's the famous 'Piano Staircase' project. In this paper, we generalize the 3 basic principles included in Fun Computing theory applicable to the clothing design method for children with ASD.

- Fun computing is a form of entertainment that use comprehensive technology and existing knowledge.
- Fun computing has the ability of motivating children with ASD to engage in more physical activity or perform onerous work in a fun manner.
- Fun computing can be widely used only for entertainment but also for physical training, psychotherapy and moral consciousness improvement.

2.2. Human-fun (Hu-fun) Analysis

In this paper, we combine Fun Computing theory and Su-field Analysis to present Human-fun (Hu-fun) analysis model for clothing design, S1 in Su-field Analysis is defined as human and S2 in Su-field Analysis is defined as clothing element. Figure 2 shows the basic model of Hu-fun Analysis.

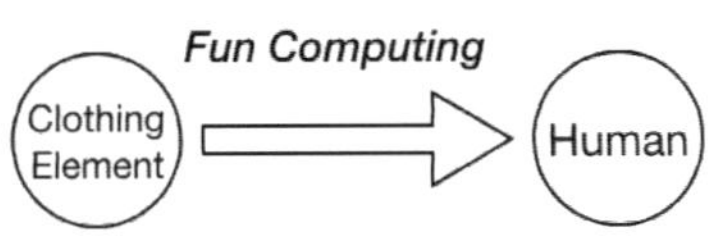

Figure 2. The basic model of Hu-fun Analysis

There are four basic principles included in Hu-fun Analysis:

1. Effective complete clothing design system for children with ASD contains children, clothing elements and fun computing.
2. Incomplete clothing design system requires completion or a new clothing design system.
3. Ineffective complete clothing design system requires improvement to create the desired clothing.
4. Negative complete clothing design system requires the elimination of the negative effect.

2.3. Making a model

The effect of clothing could be on children from the output of the fun computing. The term fun computing is used in the broadest sense, including the fields of physics and psychology. A complete clothing design model for children with ASD is a triad of human, clothing elements and fun computing.

The innovative problem is modeled to show the relationships among clothing elements, human and the fun computing. Complex systems can be modeled by multiple, connected Hu-fun Models.

There are four steps to follow in making the Hu-fun Model:

1. Identify the clothing elements. Clothing elements comprise clothing which are either acting upon human or is within substance 2 as a system.
2. Construct the model. After completing these two steps, stop to assess the completeness and effectiveness of the system. If some element is missing, try to identify what it is or find the substitute.
3. Consider design solutions from the General Solutions.
4. Develop a concept to support the solution. In following Steps 3 and 4, activity shifts to other knowledge-based tools.

2.4. Analysis nomenclature

In Hu-fun Analysis model, the human is the recipient of the system action, in this paper, human is children with ASD. Clothing element is the means by which fun can be achieved. We summarize clothing elements as:

- (St) — Style design including outer contour, inner structure and partial
- (Co) — Colour
- (Fa) — Fabric
- (Pa) — Pattern
- (Eq) —Equipment

There are six different connecting relationships between the clothing element and human expressed by different symbols and shown in Figure 3.

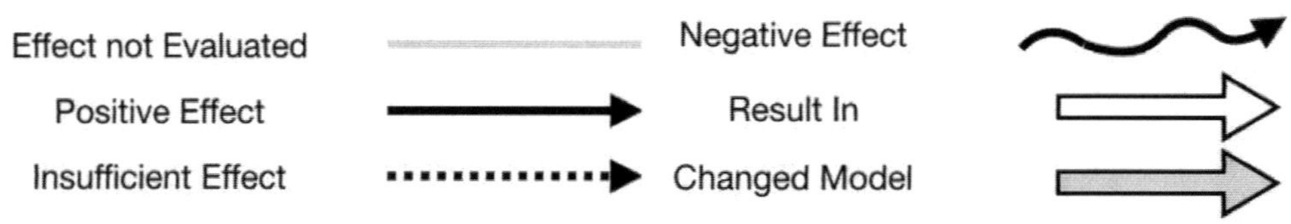

Figure 3. Six relationships between the clothing element and human

2.5. General solutions

There are three general solutions in Hu-fun Analysis based on the 7 general solutions in Su-field Analysis [11].

- General Solution 1: Complete an Incomplete Hu-fun Model

Complete a Hu-fun model if any of its three components is missing.

- General Solution 2: Modify Clothing Elements to Eliminate or Reduce Negative Impact

Designers can change internally or externally and temporarily or permanently the characteristics of clothing elements to eliminate or reduce the negative impact on the human.

- General Solution 3: Expand Existing Hu-fun Model to a Chain

Expand the existing Hu-fun model to a chain by introducing a new Sub-Hu-fun model to the system.

3. Data collection and analysis

In this study, we use a control experiment to demonstrates the effectiveness of this method by measuring idea quality and quantity.

IOP Conf. Series: Materials Science and Engineering **520** (2019) 012019 doi:10.1088/1757-899X/520/1/012019

3.1. Participants

In this experiment, participants consisted of 8 students pursuing the master degree of clothing design and engineering in the same university, they all completed the basic clothing design courses and got the credits.

3.2. Analysis of idea quality

In this study, we adopt the method presented by Shah, Smith, and Vargas-Hernandez [13]. In this method, idea quality can be sufficiently estimated even though there is not enough quantitative information to do formal analysis at the concept stage. In addition, this method adds all the quality scores for all the alternatives to achieve the total score for the set. As a result, the idea quality is defined as:

$$M = \sum_{j=1}^{m} f_j \sum_{k=1}^{2} S_{jk} p_k \Big/ n * \sum_{j=1}^{m} f_j \quad (1)$$

In this equation, S_{jk} is the score for quality for function j at stage k; m is the total number of functions; f_j is the weight for function j; p_k is the weight for stage k. The denominator is for normalizing to a scale of 10.

3.3. Experiment process

The 8 participants were asked to design clothing for children with ASD within half an hour. After the first design process, we introduced the novel clothing design method presented in this paper to the 8 participants and asked them to design clothing for children with ASD again within the same time. We define the 8 participants in the first design process is Group A and in the second design process is Group B.

3.4. Experiment results

Finally, we got all the design ideas from the 8 participants and and evaluated them by measuring their quality and quantity. Table 1 shows the idea quality and quantity of each participant.

Table 1. Idea quality and quantity of each participant

Group A	Idea Quality	Idea Quantity	Group B	Idea Quality	Idea Quantity
A1	2.54	1	B1	4.56	2
A2	2.12	0	B2	5.58	6
A3	3.41	1	B3	3.67	2
A4	1.89	1	B4	4.52	2
A5	3.87	2	B5	4.06	3
A6	2.46	1	B6	5.79	4
A7	4.12	2	B7	4.63	3
A8	3.16	1	B8	4.58	2

Then we analyzed these data with Paired-Samples T-test, Table 2 shows the result.

Table 2. The result of the Paired Sample T-test.

Group	Idea Quality $\bar{X}$	Idea Quality SD	Sig	Idea Quantity $\bar{X}$	Idea Quantity SD	Sig
A	2.95	0.082	0.08	1.125	0.640	0.022
B	4.67	0.71	<0.05	3	1.414	<0.05

Both the result of idea quality and idea quantity are significant, it can indicate that the novel method presented in this paper is an effective tool in designing clothing for children with ASD as it has the ability to promote participants' idea quality and idea quantity.

4. Discussion

Su-field analysis is an analytical tool in TRIZ theory which aims to solve technical problems. In the design method proposed in this paper, 'fun computing', 'children with ASD' and 'clothing element' are implanted into the basic model, and 'fun computing' is selected as a targeted way. By introducing a second substance or field to the original, the field is reconstructed, destroyed and transformed to achieve the effect of clothing on the autistic children's experience in a fun atmosphere, the effect of adjuvant treatment is then achieved. Through the comparative experiment, the new design method has improved the idea quality and idea quantity of designers, thus we can conclude that the new method plays a positive role in the design process. At present, the design of clothing for children with ASD is limited to the design of functions and styles, ignoring the factors of fun emotions [14]. However, compared with technical systems, the fun emotions of children with ASD is a more complicated system, so future research should be focused on analyzing, understanding, and capturing the fun emotions of children with ASD, as well as to achieve a more effective treatment for children with ASD under the results of emotion and behavior analysis [15].

5. Conclusion

In this article, we present a novel method including Hu-fun Analysis to empower designers in conceiving effective clothing design making children with ASD more interactive. In addition, we conduct a control experiment to demonstrates the effectiveness of this method by measuring idea quality and quantity. In future research, human experiments on observing the behavior of children with ASD with clothing designed by the method are most needed to explore, as well as the contextual or process factors that condition the effective use of the method.

References

[1] Rutter, M. (1978). Diagnosis and definition of childhood autism. Journal of autism and childhood schizophrenia, **8**(2), 139-161.

[2] Baron-Cohen, S., Wheelwright, S., Burtenshaw, A., & Hobson, E. (2007). Mathematical talent is linked to autism. Human nature, **18**(2), 125-131.

[3] Bauminger, N., & Kasari, C. (2000). Loneliness and friendship in high-functioningchildren with autism. Child Development, **71**, 447–456.

[4] Hanley, G. P., Iwata, B. A., & Thompson, R. H. (2001). Reinforcement schedule thinning following treatment with functional communication training. Journal of Applied Behavior Analysis, **34**(1), 17-38.

[5] Lovaas, O. I. (1987). Behavioral treatment and normal educational and intellectual functioning in young autistic children. Journal of consulting and clinical psychology, **55**(1), 3.

[6] Bailey, K. J. (2001). Social competence of children with autism classified as best-outcome following behavior analytic treatment. Unpublished doctoral dissertation. Washington State University.

[7] Weiss, M. J., & Harris, S. L. (2001). Teaching social skills to people with autism. Behavior modification, **25**(5), 785-802.

[8] Terninko, J. (2000). Su-field analysis. TRIZ Journal, **2**, 23-29.

[9] Alstott, J., Triulzi, G., Yan, B., & Luo, J. (2017). Mapping technology space by normalizing patent networks. Scientometrics, **110**(1), 443-479.

[10] Terninko, J., Zusman, A., & Zlotin, B. (1998). *Systematic innovation: an introduction to TRIZ (theory of inventive problem solving)*. CRC press.

[11] Mao, X., & Zhang, X. (2007). Generalized solutions for Su-Field analysis. TRIZ Journal.

[12] Miyata, K. (2012). Fun Computing. In Virtual Reality-Human Computer Interaction. InTech.

IOP Conf. Series: Materials Science and Engineering **520** (2019) 012019 doi:10.1088/1757-899X/520/1/012019

[13] Shah, J. J., Smith, S. M., & Vargas-Hernandez, N. (2003). Metrics for measuring ideation effectiveness. Design studies, **24**(2), 111-134.
[14] Dillon, R. (2010). *On the Way to Fun: an emotion-based approach to successful game design.* AK Peters/CRC Press.
[15] Desmet, P. (2018). Measuring emotion: Development and application of an instrument to measure emotional responses to products. In Funology **2** (pp. 391-404). Springer, Cham.

IOP Conf. Series: Materials Science and Engineering **520** (2019) 012020 doi:10.1088/1757-899X/520/1/012020

The Design of Hemiplegia Rehabilitation Equipment Based on Contextual Inquiry

Jiabao Liang[1], Bo Yan[1,a], Chenming Yang[1] and Huaiguang Yang[1]

[1]Department of Industrial Design, Hohai University, Changzhou, Jiangsu, 213022, China

[a] Corresponding author：20031640@hhu.edu.cn

Abstract. Rehabilitation equipment is the most important hardware facilities in recent medical field. The treatment methods led by medical rehabilitation instruments have been widely promoted. Besides, the study of elderlies' hemiplegia rehabilitation has become hot issues. However, the function of rehabilitation instrument on the market is simple. How to carry out design innovation on the rehabilitation equipment has become a difficult problem to be overcome in research. Hence, this paper clearly points out the goal that finish an improved design based on existing rehabilitation equipment, which is study the elderlies' hemiplegia rehabilitation by Contextual Inquiry and observation. According the study, we arranged results of key events and five working models. Then, we did an improved design and got a conclusion by the study results.

1. Introduction

1.1. Research background and motivation

With the development of society and improvements of electronic technology, People's rehabilitation therapy has changed from simple rest and medication to many aspects of rehabilitation treatment and methods. The use of instruments for rehabilitation plays an increasingly important role in modern medical treatment. In recent years, the medical rehabilitation instrument as the leading treatment method has been widely promoted with the development of science and technology, hence, apply the technology to help people recover health has become the focus of current medical work.

Central nervous system damage leads to central spasm, and the regulation of high-level motor regulation inhibits the regulation of the low-level motor center of the spinal cord, leading to the complete release of the primitive functions of the lower-level central nervous system. The brain cannot regulate it, and the influence of the external environment causes the motor cycle to be constantly excited. The affected parts of the organs cause increased muscle tone, muscle spasm, joint contracture and stiffness [1]. Due to the damage of the central nervous system, the reflection transmission is abnormal. The body's motor coordination control ability is lost, and the sense of ability to the outside environment is lost, so that the balance function of the human body is affected, resulting in sitting and standing instability.

Rehabilitation equipment is specific equipment in medical field, it belongs to comprehensive technology and plays a vital role in diagnosis, treatment and recovery. Besides, rehabilitation equipment also has positive effects on curing physical illness caused by various reasons as an adjuvant therapy. However, with the increase in the number of human diseases and growth of drug resistance, medical rehabilitation equipment technology has shown a backward trend. Innovating it has also become the focus of modern research. The main function of rehabilitation equipment is to maintain joint mobility, enhance muscle strength, muscle endurance and coordination [2].

Excellent rehabilitation instrument design is beneficial to improve the blood status of patients with hemiplegia in neurosurgery, improve motor function and daily living [3-5].

1.2. Research purpose
Improving the design of the lower limb rehabilitation training equipment for the elderly based on the existing products.

1.3. Research problems
This paper firstly hopes to find out the use groups and using principle of the rehabilitation equipment. Then, the paper attempts to use Contextual Inquiry [6,7] to understand why rehabilitation equipment has effect on the patients' recovery. Then observe that during the patient's use of the instrument, what state the family is. Last, the research wanders whether the patients can finish the training independently without families' companion.

2. Research methods
Using Contextual Inquiry (Holtzblatt & Jones, 1993) which performs the observation and interview at the user's workplace simultaneously [8]. Getting context by the method can help design the product which support user's daily activities. It is easy to find details and difficulties during the user's working time.

This paper observed and interviewed 15 patients with hemiplegia, 10 patient's family and 5 doctors, noted the records. We interviewed 30 interviewees 30 minutes each person in site. Questions covering 5 research directions, each direction with 4 to 6 questions on site. Interviewer recorded contexts and data by noting and recording. Observation focused on the patient's general training time length, training methods, degree of difficulty and security status on rehabilitation equipment. Contextual Inquiry focused on the security problems in equipment, family, doctors and surroundings during the patient's training process.

Interviewers recorded on sites and completed records after interview. Then, numbering each interviewee and making succinct conclusion. The recorded contexts were classified and researched in 5 working models [9].

2.1. Five working models
The Cultural Model reveals the harmonious relationship between patients and families, patients and doctors, families and doctors.

The Tool Model shows how to use the rehabilitation equipment, whether the equipment is safety and convenient for the patients to operate. Whether the recovery achieves the desired effect by using the rehabilitation equipment.

The Physical Model shows the patient with equipment in physical environment and the safety issues surrounding the patient in using the equipment.

The Task Model shows the patient in performing the rehabilitation process, whether each step can achieve the desired purpose of training, recovering and using.

The flow Model depicts the rehabilitation equipment that is safe and smooth to perform each necessarily communication and cooperation in safety and convenient aspects.

3. Research results

3.1. Key event results
According the mentioned methods, we interviewed 30 research objects (15 patients,10 families and 5 doctors), got the related interview records. Then, organized and analysis the records. Finally, the key

events and key time which got from the interview were classification recorded in the form of tables. Results are shown in Table 1 Key events results.

Table 1 Key event results

Key Event	Description	Findings	Advice
Patients can't lean on the safe belt	The safe belt can't provide enough security for patients, so they prefer grip the table board to lean on safe belt.	Because of design problems, safe belt for ordinary elastic cloth can't provide security for patients.	Offering a foothold or protection for patients.
When the patient uses it, he must be accompanied at all times.	The safety of the device is low, the patient and family cannot be assured.	The equipment is no able to give the patient enough support and fixation.	The equipment can be used separately for patients
Patients grip the table board tightly.	Patients are afraid of falling when using the device, and it is very laborious to grab the board.	Desktop design has certain defects, just a smooth board without any supporting part.	Providing a supporting part on the desktop for user.
User's hands lean on the table board constantly, it is an uncomfortable posture.	Low safety factor leads to the uncomfortable operation.	The table board is fixed on the stable angle, patients are unable to adjust the table board in a comfortable angle.	Providing certain support for patients to finish training.
Patients' upper body leaning forward because of insufficient support, families have to help them adjust posture.	It's hard for patients with hemiplegia stand without back support.	Because of the patients with hemiplegia, it is necessary to provide assistant part for them to finish training.	The equipment with supporting part to stable the patients' upper body when they training on it.
When adjusting the pedal angle, the patient needs to leave the instrument first. Besides, angle choice is limited.	The knob which adjust the angle was designed on the bottom of the equipment, most patients has the difficulty to twist the knob.	Jagged design limited the angle choice, it leads to the discontinuous angle choice, the knob designed in an unreasonable position.	Optimizing the adjusting system, positioning the knob on the reasonable position where user can touch when they standing on the pedal.

3.2. Five Working Models

Model 1: The Physical Model

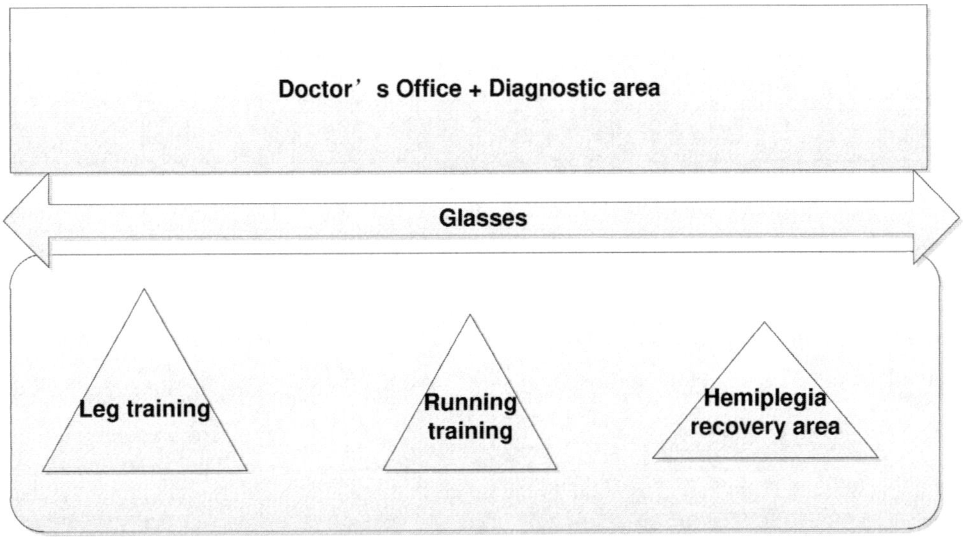

Figure 1. The Physical Models

As Figure 1 shown, the glasses of doctors' office are fully transparent. This design is convenient for doctors to observe the patients' recovery state and training process.

Model 2: Information\ideas Flow Model (Between interested parties)

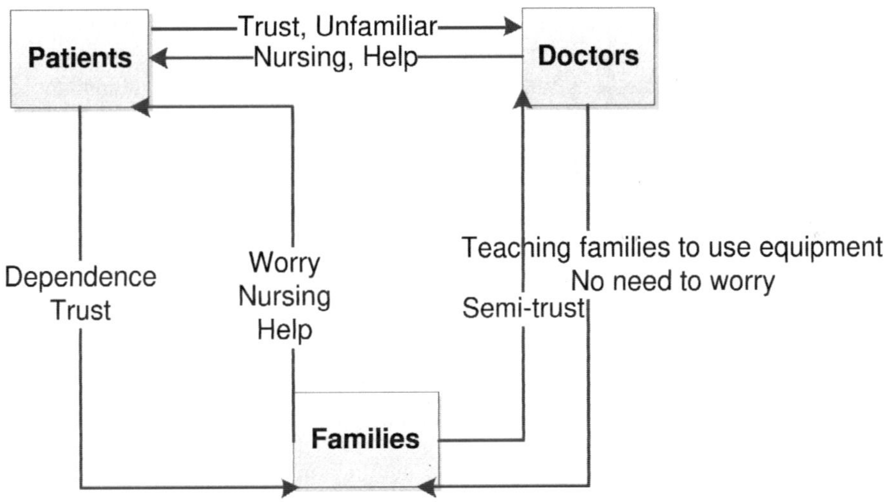

Figure 2. Information\ideas Flow Model

As Figure 2 shown, the patients trust their families certainly. Families will help them finish the recovery training, besides, families also accompany the patients all the time to avoid accident happened during the rehabilitation.

Doctors usually taught families to use the equipment, as how to adjust the equipment. Just as the Physical Model shown, doctors can observe the patients' training process, hence, families don't worry too much about the security problems.

Most patients semi-trust doctors, they trust doctors because of their professional guidance, however they are doubt about whether doctors can find they troubles in time when doctors in busy. On one hand, doctor guide patients to recover from illness, patients trust doctors' profession. But compared to patients' families, doctors are unacquainted, if their relations absence, elderly patients sometimes cover up their problems because they afraid to trouble doctors. This phenomenon results from elderlies' self-respect.

Model 3: The Task Model

Table 2 The Task Model

Action	Context	Steps	Problems
Stand on the equipment	Standing on the equipment with relations' help.	Stand on the equipment Strap the safe belts Adjust the pedals	it is trouble to adjust the pedal angle.
Recovery	Stand on the pedal	Standing on the pedal for about 30 mins.	Lack of security Leaning on table board is tired Boring process The distance between two table boards is too close.
Leave the equipment	Leaving the equipment with relations' help.	Release the safe belt	

As Table 2 shown, this paper divides The Task Model into 3 parts, stand on the equipment, recovery and leave the equipment. We found that relations accompany the patients all the time. First, relations have to help patients stand on the equipment, strapping the safe belt on their back. Second, patients have to repeat the stand and leave the equipment process because of the knob designed on the bottom of the equipment. This tedious operation result that patients seldom adjust the pedal angle which is suitable for themselves. Third, the oversimplified safe belt cannot provide enough support for patients, it results them afraid of falling from the equipment, so they have to lean on the table board tightly. This action will waste energy with doubt. Fourth, when patients standing on the equipment for recovery, they were unable to do other things, this process is boring. The distance between two table boards is too close

Model 4: The Tool Model

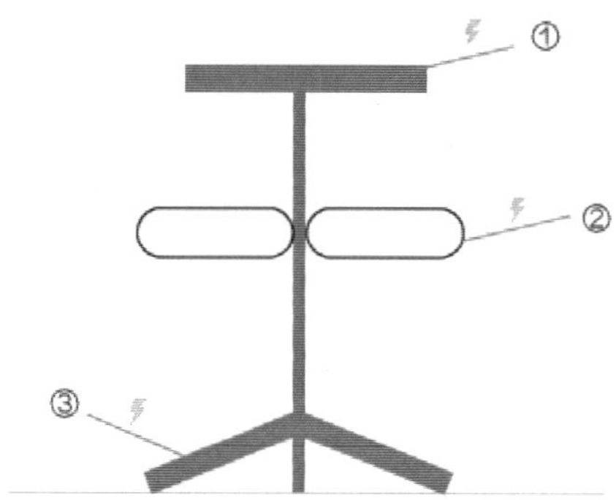

Figure 3. The Tool Model

As Figure 3 shown, this is a Schematic diagram of equipment, ①Table Board, ②Safe Belt, ③ Pedal. The distance between the two table boards is about 50 centimeters. This length is unsuitable for two patients stand face to face. Due to the areas limitation, when patients put hands on the table, there is nowhere for them to put their private good such as towel, cup and phone.

The most important problem of pedal is knob, it is unrealistic to repeat stand and leave the equipment, moreover, most people don't know how to adjust the pedal angle, they choose to accept the uncomfortable angle, it will do harm to their ankles. Block type design is unreasonable.

Model 5: The Cultural Model (Attitude to key events)

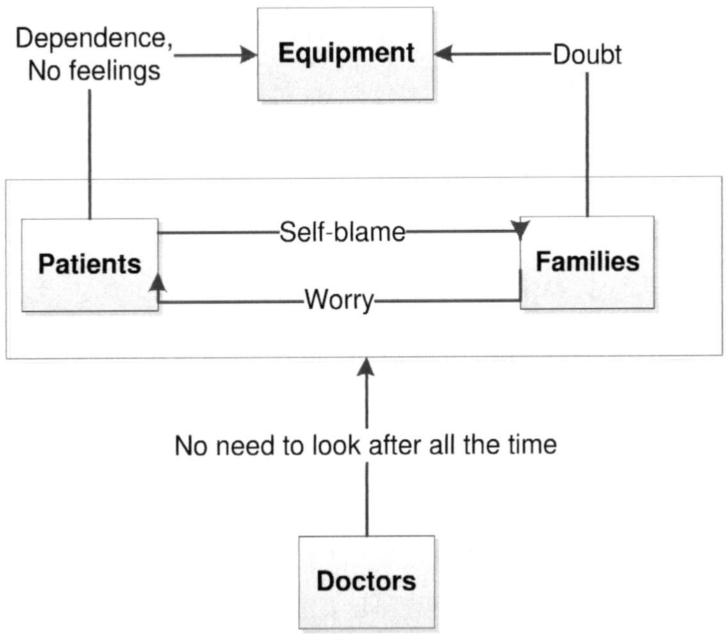

Figure 4. The Cultural Model

According to observation, interview and research, the Cultural Model is shown in Figure 4. The research found that patients actually know little about the equipment they used. They rely on the equipment because they think it will help them recover from the illness, however, they have no trust on the equipment. They seldom have positive attitude on the equipment. The ideal situation is that patients have a positive relation with the equipment, then, they are willing to recover on it.

Elderlies worry they will be burden of their children, they hope to prove themselves. Under these circumstances, the elderly patients with miplegia always have large pressure. They are more likely to feel that they are the burden of others because of the illness. The problem will be infinitely magnified if they found they lose the ability to training independently. Meanwhile, pressure and worries pour out.

From the perspective of family, they will never leave the elderly patients training alone. Doctors may not take care of each patient carefully due to the number of patients.

3.3. Analysis and Summary

Analyzing the statistical results come from tables and models

1. Patients worry about the equipment's security issues, so a more reasonable structure will bring them more feeling of security when they use the equipment.
2. The table board on the equipment needs to assembling handgrip to make sure balance of patients' body.
3. The design of pedals should consider patients' personal demands. Optimizing the adjustment method is necessary.
4. Taking account in the long time on the equipment causes boring, certain entertainment facilities are able to attach in the equipment, but it is not necessary.

In summary, improved design existing in the redesign of safe belt on back, redesign of table board and the structure design of pedal.

4. Design Innovation

4.1. Attaching pads on the back and the legs.

Differences: before design, Lack of security.
After Design, increasing security.

4.2. Adjustable height handrail and leg braces
Differences, before design: patients' upper body leaning forward and hands griping the table board because of insufficient support.
After design, providing a handrail for patients to put their hands and they can adjust the height according to their personal requirements. The height of whole equipment could be adjusted to cater to patients' height.

4.3. Changing the material of table into more soft material and adding borders around the table.
Differences: before design, wooden is hard for elderly patients and smooth surface personal belongings are easy to slip from the smooth surface.
After design, increasing the comforts of the table, borders protect the belongings from slipping.

4.4. Adjusting pedal from manual adjustment to automatic adjustment, controlled by interface.
Differences: before design, manual adjustment is complicated, patients can't finish the action independently. Besides, angle selection is not smart enough.
After design, automatic adjustment is easier for patients to select their suitable angle, they can stand on the pedal and adjust the pedal angle by operating on interface. This design makes them choose angle directly and quickly. They can also adjust the angle by themselves.

4.5. Adding a display to the equipment.
Differences: before design, long time training brings boring and discomfiture.
After design, three advantages. First, entertainment eliminates boring during the training, it makes the training become more easily. Second, timing function, this will remind patients to pay attention to training time, creating a normative training plan. Third, adjusting angle precisely can be realized by controlling the virtual button on display. The value shows more intuitive feelings.

4.6. Adding emergency button.
Difference: before design, doctors and relations have to pay attention on the patients all the time, they constantly asking patients for feedback.
After design, connected with doctors' computers, doctors can receive the information at any time. Emergency button can make patients call for help immediately when they feel uncomfortable during training.

5. Conclusion
This paper is based on the increasing number of elderly patients with hemiplegia and the important of using medical equipment products for rehabilitation to interview by Contextual Inquiry. According to the relevant information about the ankle rehabilitation equipment, we confirm the Human Factor (Users) is patients with hemiplegia, the research purpose is designing an ankle rehabilitation equipment which is more convenient and efficient for patients. From God's perspective, we set ideal state and use it to minus the present state. In this way, the preliminary questions came out. After Contextual Inquiry and observation, further problems were excavated such as pedal angle problem, safe belt problem and table board problem, etc. Last, in trust design of present product, a new ankle rehabilitation equipment was designed based on the improvements of mentioned problems.
This paper excepts the research methods which Contextual Inquiry and using God's perspective minus present state to find problems can provide a solution for other products' improved design.

Acknowledgments

This paper is written through Professor Yu-Hsiu Hung's instruction. This work is supported by the the Practical Innovation Program for College Graduates of Jiangsu Province (Grant No. SJCX18_0196) and Fundamental Research Funds for the Central Universities (Grant No. 2018B800X14)

References

[1] C.Y. Hu, Development of a hemiplegia upper limb compound exercise rehabilitation training robot, (Tsinghua University Press, 2004)
[2] X.G. Wang, H.J. Zhang, CN. Chin. J. 07, 102-104, (2014)
[3] X. Huang, H. Wu, Effe, CIPE, CN, (2008)
[4] S.J. Gong, J. Zhang, K. Wu, CN. Chin. J, 05,848-849, (2003)
[5] Y. Zhu, Y. Hu, Y. Wu, C. Jiang, W. Fan, Neur. J, 11,670-674, (2007)
[6] Blechner, M., Monaco V., Knox I., Crowley, R. S., AMIA (2003)
[7] H. Beyer, K. Holtzblatt. Contextual Design: Defining Customer-Centered Systems. (San Fransico, 1998)
[8] H. Beyer, K. Holtzblatt, Contextual Design. Interactions, 6(1), 32 - 42. (1999)
[9] Y.H. Hung, W.W.W. Lii, HFES, 52(21),1718-1722, (2008)

IOP Conf. Series: Materials Science and Engineering **520** (2019) 012021 doi:10.1088/1757-899X/520/1/012021

Sustainable Open Public Spaces: Place Making Strategy for the Breakwater "Al Kaser", Abu Dhabi

M. Kaftangui[1,a], B. Welling[1], H Masalmeh[1] and Y Anbar[1]

[1]Architecture Department, Abu Dhabi University, Al Ain road, P.O.Box 59911, United Arab Emirates.

[a] mohamed.elkaftangui@adu.ac.ae

Abstract. The Abu Dhabi Vision 2030 revolves around sustainable development and seeks to raise the standard of living across the Emirate. Parks are one of the important tools to generate Abu Dhabi vision, where it acts as the collection point of all varieties of people. Aware of the important demand of proper Parks, Abu Dhabi Municipality, is seeking to adopt park's facelift to remedy to some weaknesses appearing in a majority of Abu Dhabi parks to raise the quality and the liveability of these open urban spaces. This paper will study a potential sustainable public open space in the country, the case of Abu Dhabi Breakwater (Al Kaser). This space holds a great historical value and has an exceptional location where individuals can enjoy the skyline of the city over the water edge, taking in the picturesque views and cultural fascinations. The aim of this paper is to propose a new framework for developing a vibrant public open space including new and unconventional cultural activities in the area of Abu Dhabi. By presenting a literature review supported by case studies, place making strategies and a site analysis of Al Kaser, the paper reaches a sustainable retrofit conceptual Master Plan proposal of the breakwater. Taking into consideration the social, economic and environmental factors of the proposed design is the key to a sustainable outcome.

1. Introduction

Al Kaser is an urban attraction that spun from historical purposes, it served previously as a breakwater to block waves, therefore, protect the Corniche of Abu Dhabi. Formed out of rocks and solid foundations Al Kaser consisted of a single pedestrian walkway where fishermen would make use of the coastline protection. Today, Al Kaser is an attraction primarily for the youth and adults where they cruise around the location, take in the picturesque views and cultural attractions. This site has come a long way since being formed in the 1960's to holding events, activities and cultural purposes in the modern world.

Public spaces are all around us acting as an open-air living room, like an outdoor leisure centre which means they should be plentiful, accessible, unique and well designed. Properly designed these spaces can bring communities together, act as meeting places and help develop social ties that as of late have been disappearing in many urban areas.

The Abu Dhabi Vision 2030 revolves around sustainable development and seeks to raise the standard of living across the Emirate [1], because there is a big need for public open spaces in Abu Dhabi even with the hot humid weather condition during the summer. Parks and playgrounds are the main public open spaces in Abu Dhabi, however some have difficult accessibility are very inaccessible due to their location next to main roads, these parks become so crowded during the weekend that the designated car park spaces fill up quickly and others are left to park on the hard shoulder. In a recent survey, two thirds of households said that the emirate lacked proper public spaces [2].

This study is conducted on a premium public space called the Breakwater, shown below in 'Figure 1', better known to locals as 'Al Kaser' this site is located in the emirate of Abu Dhabi, it is a public open space with great views and is usually quite busy, however, we will be designing a place making strategy by addressing its weakness highlighted in the third paragraph.

Figure 1. Al Kaser Map & Location, Abu Dhabi, UAE. (Source: Google Map)

2. Historical Overview

Al Kaser is a man-made breakwater barrier, it was established as a strong barrier for breaking high waves that could have led to several problems, particularly the destruction of what was newly built on mainland Abu Dhabi at the time. A number of large rocks, where loaded onto trucks and transported from Ras Al Khaimah to Abu Dhabi. After its completion, local fishermen took a particular interest to the site and used it for leisure, where others used it as a promenade as shown in 'figure 2'. Over the course of time this site has been developed and now has great purpose for Abu Dhabi tourism [3].

Figure 2. Al Kaser in 1960, Abu Dhabi, UAE

3. Site Problems

The existing typology scheme of Al Kaser faces many issues such as the lack of activities, limited pedestrian paths and site accessibility, inefficient land use, lack of urban furniture, obstructions blocking the best views of the site and restricted dining options and many aspects that are not used to their best potential as shown below in 'figure 3'.

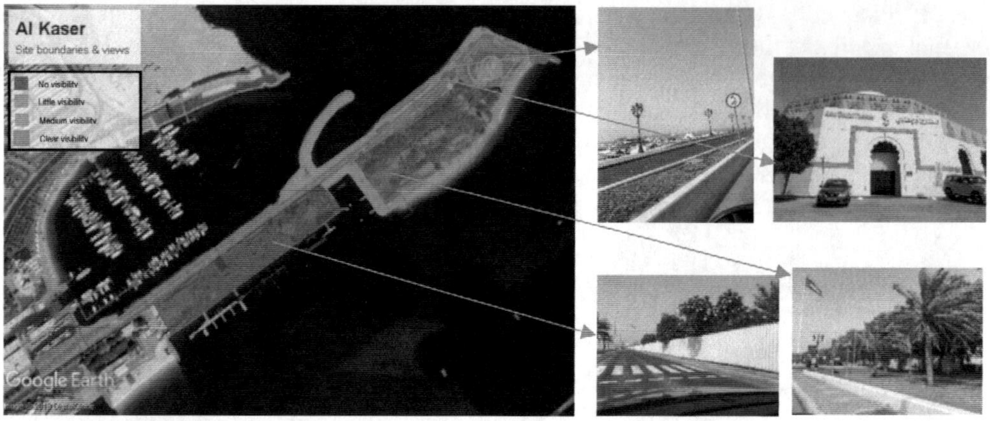

Figure 3. Al Kaser View Restrictions. (Source: Google Map & Author)

The following table 'table 1' summarizes the main issues that are facing Al Kaser and the upgrade that was proposed in this research.

Table 1. Al Kaser problems and proposed solutions. (Source: Author)

Problem	Proposed Solution
Reinforced straight path, one circulation method	Smother pathway to enhance visibility, and interaction to provide a dynamic and engaging experience.
One access road for cars/ limited pedestrian path	No car access to promote walkability and dividing the pathway to pedestrian, biking and jogging track. Allowing only Eco-cars/ golf carts on site. Providing an appropriate parking area before accessing Al Kaser.
Blocked views from buildings/ fences	Removing walls and fences to allow visual and physical access through buildings without restrictions.
Private Yacht Club on site	Introduce a Public Yacht Club without fence enclosure that includes water activities and a Public beach.
Negative Interface of Cafes	Cafes open to view from different sides allowing interaction with the surrounding and the users.
Limited seating areas	Additional shaded seating areas and rethinking the urban furniture in site.
Visual Connectivity Constraints	Remove any fences/walls around existing buildings and maximize pedestrian exposure from surrounding areas.
Accessibility	Proposing new pathways to improve access to the public space and reach site from different directions.
Spatial Organization	Exploring spaces around site to increase interaction.
Restricted Views	Extending observation points and removing blocking walls and fences.

4. Proposal

In this paper we propose a conceptual Master plan strategy for a public open space to bridge the gap in the urban fabric of the city between urbanism and nature while promoting integration and socialization across different lines of cultures [4]. The proposal intends to promote the space as a meeting point which will connect Abu Dhabi's diverse community into a natural hub away from their daily life style. This space will also provide comfortable and productive space for people to engage in various activities.

4.1 Sustainable Design Concept

The design concept behind this research proposal is inspired by the natural features of the site. The fluidity of the water over The Breakwater has been the main inspiration to create a smooth and seamless pattern that achieves a dynamic circulation spine around the themed areas. A series of smooth curves will converge inward and extends outwards synergizing the pier within the sea trying to enhance the connectivity to nature.

The concept of fluidity enhances the integration between public and private areas as it allows users to walk through and around buildings and spaces on site on smoother walkways without boundaries that block their views and limits their accessibility [5].

The proposed urban space also introduces new activities on site to follow the concept of fluidity [6]. The variety of activity on site allows the space to become more dynamic and attractive to the users. The activities are shown in 'figure 4' below.

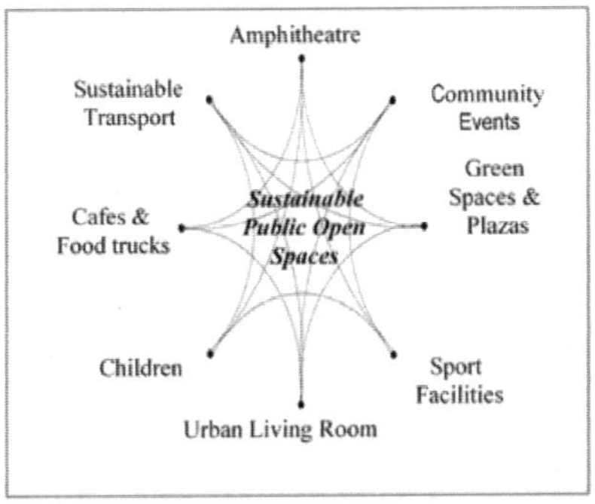

Figure 4. Al Kaser Fluidity of Activities within the Public Open Spaces. (Source: Author)

4.2. Accessibility, Circulation & Zoning

The current plan of Al Kaser includes one main access point and a linear path for circulation around the urban space 'figure 5'. This linear path is mainly used for car access with limited space for pedestrian walkways on both sides of this path. There is no space for other activities, such as biking and exercising, as the current pedestrian walk way serves those functions as well. Having this linear path limits the visual and physical connection of the urban space, therefore; proposing a more dynamic primary path that passes through the urban space with secondary paths extending from it and adding more access points would solve this issue 'figure 6'.

The addition of curved circulation along with curved elements enhances fluidity of the space giving a more exciting experience for the users. The curved elements extending from the site will serve as observation points for the views of Abu Dhabi's skyline.

The existing urban space is divided into six main zones: Culture, Parks, Sports, Community, Parking and Café's shown in 'figure 7'. The new proposed zoning rearranges these zones for a more successful urban space and includes two new zones: Observation Points and Plazas 'figure 8'.

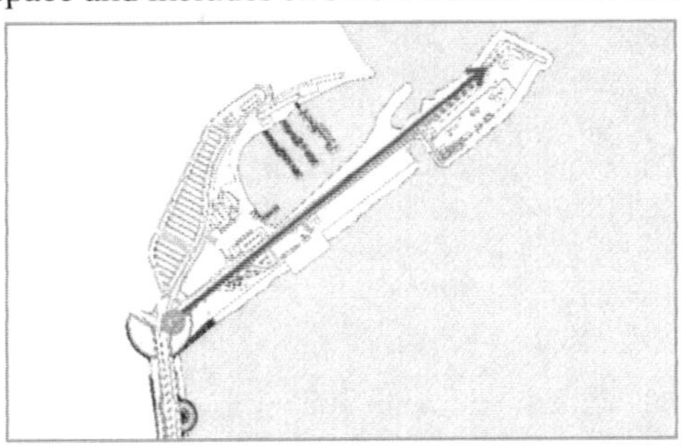

Figure 5. Al Kaser Fluidity of Activities within the Public Open Spaces (Source: Author)

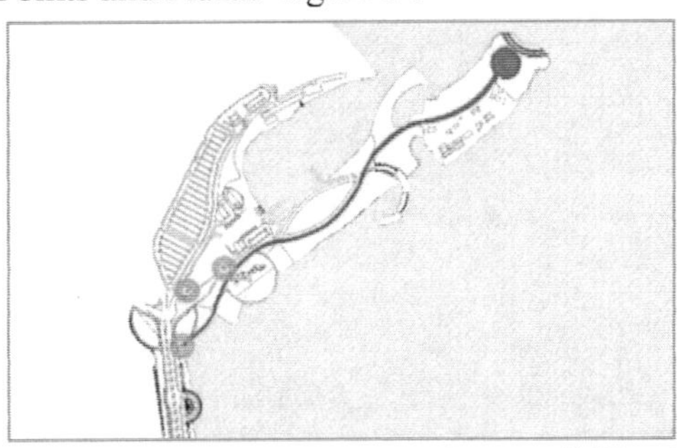

Figure 6. Al Kaser Proposed Accessibility & Circulation with new elements (Source: Author)

IOP Conf. Series: Materials Science and Engineering **520** (2019) 012021 doi:10.1088/1757-899X/520/1/012021

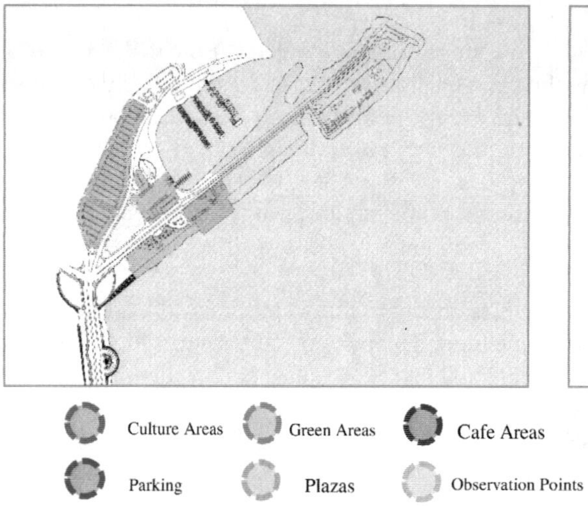

Figure 7. Al Kaser Existing Zoning (Source: Author)

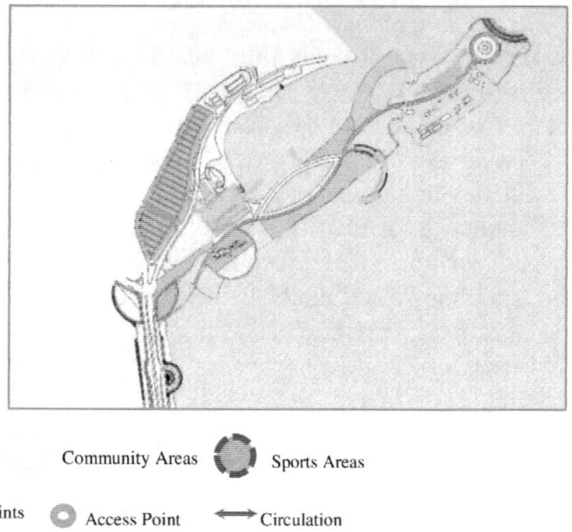

Figure 8. Al Kaser Proposed Zoning (Source: Author)

5. Master Plan Proposal

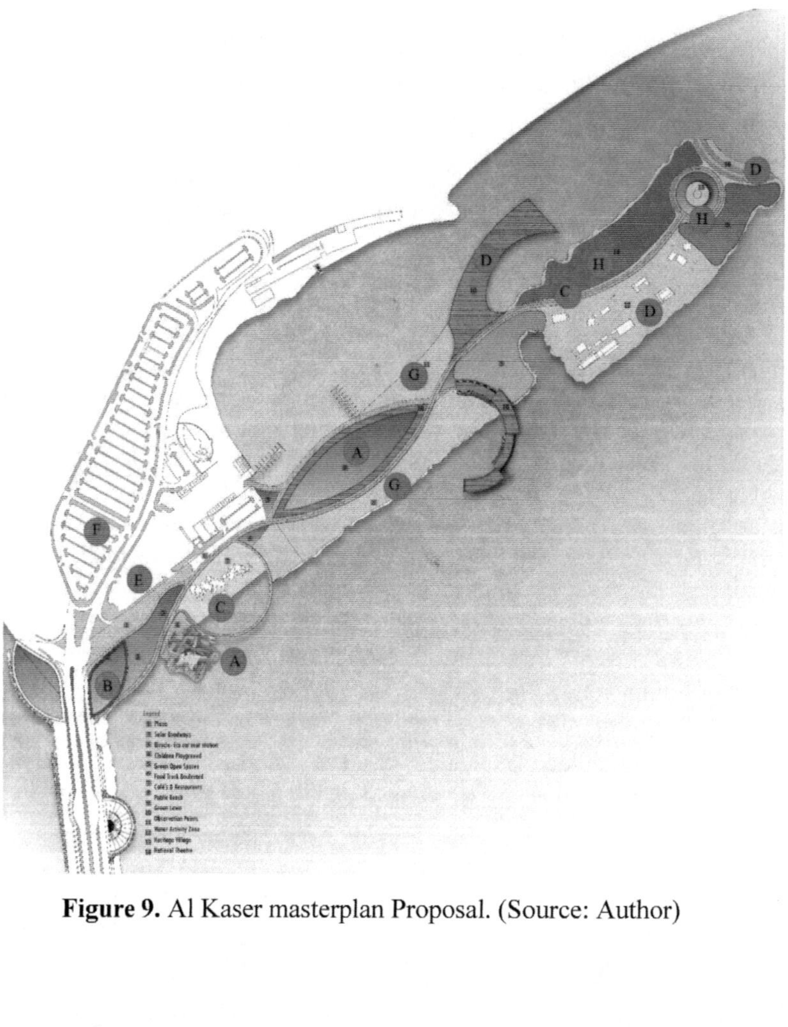

Figure 9. Al Kaser masterplan Proposal. (Source: Author)

IOP Conf. Series: Materials Science and Engineering **520** (2019) 012021 doi:10.1088/1757-899X/520/1/012021

5.1 Master Plan Features

The Master Plan introduces various new features to the open space 'figure 9'. Starting with accessibility, the plan introduces three access points to the urban public space which then have pathways divided into pedestrian, bicycle and jogging tracks. Plazas are integrated into the new plan to hold art pieces and sculptures from the UAE's culture and the natural context of the site. As for vegetation, it is important to use indigenous plants for landscaping; meaning that it is native to the UAE or the region. It is also important to organize vegetation in a way that doesn't block views but instead provide shade and beauty to the site.

In addition to vegetation, solar umbrellas and canopies are proposed as shading elements on site. They are arranged in a way that helps preserve the views while providing comfort for users in hot seasons. These devices include fabric and transparent material as well as solar panels that are used to collect energy and power some of the site features.

The space includes shuttle services as a transportation alternative to cars. These shuttle services are in shape of eco-cars or golf carts that can transport users and materials on site. They are powered by solar energy that is collected using the shading devices.

Solar panelled LED lighting can be used in features and displays in Al Kaser to create a great night time atmosphere as that's when the site is usually the busiest. Light patterns and light fixtures will add a liveliness to the area and encourage more people to use the public open space built in a sustainable manner. Street lights must use LED light bulbs as they have significant advantages over other forms of lighting. [7]

In order to make the site sustainable, we must take special precautions even down to the footpaths. Kinetic pavements would transform Al Kaser, which is currently a concrete and brick paved site, kinetic pavements can be made using 95% recycled tires that can flex up to 5mm when walked on. These can also produce electricity of 8 watts that can be stored in batteries to power street lights and other electrical items. These paths are very strong and withstand harsh climatic conditions which would make them perfect for Al Kaser.

To mark the road for pedestrians, cyclists and eco cars solar roadway panels can be used; these are intelligent panels that would replace paint that can be toxic and less aesthetically pleasing. These panels can be customized and can even warn pedestrians when eco cars are incoming.

Around the restaurants an important area of vegetation and grass are provided to allow an appropriate natural rainfall drainage system during floods seasons, small paths of gravel can be laid down to direct people to the cafes and restaurants.

5.2 Mater Plan Content

The Master Plan contains the following:

A. Observation Points: An additional feature that enhances the users experience to the views of Abu Dhabi Skyline through an amphitheatre and decks 'figure 10'.

B. Plazas: Community areas where people gather and interact socially as they share interests in art and culture. These spaces are equipped with shaded seated areas and solar stations with bikes and eco-cars figure 11'.

C. Green spaces: Distributed on site, includes parks and kids play areas 'figure 12'.

D. Dining Areas: Includes food trucks and cafes distributed on site 'figure 13'.

E. Community Centre: A space where the community connect and engage in indoors events during the hot seasons. It includes a gallery space, sports facilities and a theatre.

F. Parking Space: Where the car access stops and users can enter the public space either by pedestrian walk way or by renting various sustainable transportation such as eco-cars, golf carts or bicycles.

G. Water Activities: People can enjoy the waterfront by engaging in the water activities on both sides. Users can rent boats and jet skis or enjoy walks and sunbathing on the public beach.

H. Culture District: This district is one of the most important elements to the master plan, as it acts as an Urban Living room that includes the current heritage village and Abu Dhabi National Theatre.

This space holds cultural and recreational activities for people to participate in, as well as markets that include food and items from different cultures.

Figure 10. Amphitheatre that acts as an observation point towards Abu Dhabi Skyline. (Source: Author)

Figure 11. Bicycle and Eco- cars/ golf carts Solar Energy station. (Source: Author)

Figure 12. Green areas with water features. Pictures shows pedestrian pathway along with biking track and shading devices. (Source: Author)

Figure 13. Food Trucks and seating areas viewing the water front and Abu Dhabi Skyline. (Source: Author)

6. Reflexion and Conclusion

The Breakwater is one of the few places in Abu Dhabi that has a great historical value with an exceptional location viewing the skyline of the city over the water edge. Unfortunately, due to the existing typology scheme it has limited the community from visiting this place due to the different factors listed above. This paper reached a proposal that could enhance the existing public space turning it into a more sustainable and interactive public space The consideration of the social, economic and environmental factors of designing this space is the key to a sustainable outcome providing an interactive space with more activities, comfortable atmosphere and environmentally friendly features as summarized in the 'table 2' below [8].

Table 2. Sustainable aspects proposed for The Breakwater, Al Kaser in Abu Dhabi. (Source: Author)

Social Aspect	Economical Aspect	Environmental Aspect
Adding comfortable shaded seating areas	Encouraging small businesses (such as food trucks and markets)	Using renewable energy (e.g. solar energy) on features on-site
Boosting social and physical activities	Enhance tourist activities and cultural attractions	Limiting car use and depending on eco-cars.

The proposal aims to transform the breakwater into a highly interactive and safer environment for the community providing easy access for pedestrians and parking solutions that work within the existing surroundings. The plan will enhances lighting, shading, pedestrian flow, and access points into site. The new public open space aims to give a chance for visitors to have a special experience of exploration and discovery, offering a journey through a series of themed areas.

References

[1] The National. Abu Dhabi plans for better public spaces. (2011, April 06). Retrieved from The National: https://www.thenational.ae/uae/abu-dhabi-plans-for-better-public-spaces-1.433510
[2] Abu Dhabi Urban Planning Council. Plan Abu Dhabi 2030. Abu Dhabi: Abu Dhabi Urban Planning Council. (2007)
[3] Al Bayan Newspaper. Abu Dhabi Corniche A tale of history and Renaissance in 1970. (2012). retrieved from: https://www.albayan.ae/paths/storytelling/2012-11-11-1.1764320
[4] Shelton, I. Principles for Sustainable Public Open Spaces in Sandton: A Recommendation for the Johannesburg metropolitan open space system (JMOSS). (2016). Retrieved from: http://wiredspace.wits.ac.za/jspui/bitstream/10539/22348/2/Research%20 Report%20Ingrid%20Shelton%20317011.pdf
[5] Campbell, C. Visualizing Residual Spaces in a New Light: A Study for Campbell, C. Visualizing Residual Spaces in a New Light: A Study for Creating and Designing an Urban Public Space Under a Downtown San Jose Freeway Creating and Designing an Urban Public Space Under a Downtown San Jose Freeway. San José State University Urban and Regional Planning Department.(2013). Retrieved from: http://www.sjsu.edu/urbanplanning/docs/URBP298Docs/URBP%20298%20-%20Honors%20Report%20-%20Campbell.pdf
[6] Project for Public Spaces, Inc. Place making for Social Cohesion and Economic Sustainability: A training Seminar facilitated by Project for Public Spaces. Manual for Workshop: 9-11 November 2011. Project for Public Spaces, New York. (PPS) (2011)
[7] LEDs. Solar Roadways. (n.d.).Retrieved from: http://www.solarroadways.com/Specifics/LEDs
[8] Reiter, S. Correspondence between the conception principles of sustainable public spaces and the criteria of outdoor comfort. Paper presented at the 21th Conference on Passive and Low Energy Architecture, Eindhoven, The Netherlands, 19-22 September

IOP Conf. Series: Materials Science and Engineering **520** (2019) 012022 doi:10.1088/1757-899X/520/1/012022

Research on the Child-oriented Design for Community Libraries

Ting Hou[a] and Bo Gao

Nankai University Binhai College, Tianjin, China

[a] 124197808@qq.com

Abstract. Based on the reading behavior and reading psychology of children as well as the varying preferences among different ages, and with some references to the cases and the status quo of public libraries in China and the United States, a community library has been designed in this article. For the appearance, animal shapes are widely used in the design. The space is divided into different zones for different age groups. Dynamic zones are the main part of the design, which makes the space more open for communication and activities, in order to meet different needs and create a community library that really serves children. With an environment like this, the atmosphere will thus be enhanced for children's independent reading.

1. Introduction

Reading is the link of knowledge spreading and cultural communication, and also an important method for learning and mind purification. With the change of the times, books still play an important role in child development now in this information age. For children, reading is the beginning of their literature life, the formative education of art, and their cognition and experience of life and the world.

Reading has to be in a certain environment. If you want children to love reading, you have to make them love the atmosphere of reading before they read. Therefore, a space for children communication and education has to be created, and a reading place with good environment, facilities and services has to be designed.

In recent years, China has strongly pursued nationwide reading, and the government and society pay more and more attention to children's cultural life, as children are the future of the country. Some public reading zones have appeared in bookstores, which provides good places for children to read. However, those places are usually located in commercial districts where mainly provide services for adults. The spatial design and the size of facilities there are also not suitable for children. Children may also be dangerous when going out. Those are the problems that must be conquered. But how? A child-oriented community library will give a perfect solution.

2. Public library and community library

Public library is an important place for public service and knowledge spreading. It is the modern public reading space. Although started late, Chinese public libraries are continuously improving with belief of reading. The United States has a mature experience in public libraries, and is the first to establish legislation for public libraries. Community library is a branch of the public library system. It has not only the same functions with public library, but also unique charms making it appropriate for communities.[1]

2.1 Public library and community library in the United States

The United States is an early country with better results in developing public libraries.With its distinctive features, American public library works so well that has become a model of all public libraries across the globe. Public library is popular among American people for its free admission to public space and excellent services.[2]

Philanthropists and civil leaders first set up public libraries for those who wanted to read but couldn't afford to buy books. Through borrowing books, readers who were once just strangers could become close friends. The function of American public libraries can be defined as the English word "free", freedom, and free admission.

Community library is a huge part of the American public library system. American community library has formed the most mature system. A community library takes the important role as a center of entertainment, communication and education.

2.2 Public libraries and community libraries in China

The development of public library in China can be divided into three periods: the public libraries in the late 19th century, the Imperial Library of Peking during the Republican period, and the national public library system since the founding of People's Republic of China.

Public libraries in China are unevenly distributed across the nation, since most of them are located in cities. The number of public libraries in the major cities in the east is far beyond that in the west of China. For site selection, libraries are usually located near schools or government agencies, or in the commercial districts. However, such locations could only serve the upper class people.[3,4]

3. Child-oriented service

In recent years, children reading has appeared on stage. Children's libraries take a huge part of community libraries. Despite reading as the basic function for library, it is more important to give children the chance to discuss, to love reading, and for their lifelong learning.

3.1 Child-oriented service in the United States

Children's library has a long history of over 200 years in the United States, and almost every community has a children's library for children to read and learn. But even in the United States, children under 14 years old were not allowed to enter libraries, for they could make too much noise and disturb the quiet environment in libraries. However, with the change of times, Americans have realized that children could be easier to become a talented person if they receive education early. After over a century of development, children's libraries in the United States have developed a creative system of their exclusive collections, space, services and collaborative networks.[5]

First, exclusive collections. A good collection of books in a children's library is determined by the requirement of children. To meet children's requirement, the books should be fun, easy to read, and with intellectual features. In that way, children could be attracted to read, and also learn some knowledge. Another important point is to meet their interest. Children under 12 years old usually like fairy tales and craft books, while those over 12 years old may more likely to choose novels and comic books. All in all, the collection should be diversified and abundant.

Second, exclusive space. The spatial design of the building, the layout, and its ornaments should accord with children's mind. A children's library should break the traditional method of space layout. Inflexible segmentation of space has been prevented in order to increase service area. A large, open and bright reading space would be there welcoming the little readers. Color, safety and practicality should also be concerned in the indoor design. For example, wall paintings with full of children's favorite elements, ergonomic design of little tables and chairs and low shelves for children……Through the bright, colorful and enthusiastic design of the library, a lovely reading environment will thus be created for children.

Third, exclusive service. The biggest feature of services in the United States is its targeted service. Children are diversified into different age groups. New methods and contents are continuously invented in terms of different preferences of different age groups. Activities within certain age groups are also attached great importance. For example, storytelling and parent-child reading for infants, and homework assistance and summer reading for school kids.

Finally, collaborative networks. Libraries should keep in touch with schools, parents and communities, and thus form a quaternary system, so that the service will be extended. American libraries pay much attention to the double development both online and offline: online resources for children, and offline activities in the reading area. The Internet platform can be a supplementary resource for children's library.

Service is the first step of the work for library. The child-oriented service focuses on reading with joy, and cultivate children's interest in reading. The American concept of child-oriented service follows the mind of children and attracts children to library in their own will. That's the real service for children.

3.2 Child-oriented service in China

Children reading has become one of the hottest topics. In China, child-oriented service has taken its shape, but is still behind America. There are plenty of shortcomings in forms, contents and ability. Chinese libraries mainly focus on its facilities, but there are just few libraries that particularly provide service for children. The current situation of child-oriented libraries in China can be described as "strong in the south but weak in the north, and rapid in the east while slow in the west".

China has borrowed the five basic elements in child-oriented service from that in the United States, although not that perfect.

First, exclusive collections. There are rich collections in Chinese libraries. They are almost perfect in both quantity and quality. China does the best in library collection among these five elements.

Second, exclusive space. The indoor spatial designs of children's libraries are basically the same with libraries for adults. Their spatial plan, colors and functions cannot satisfy child readers. Normal libraries usually appear serious and solemn, which is just the opposite with those active children. It's not good for children's development.

Third, exclusive service. Not so many services are available in Chinese libraries. However, children are usually not attracted by a single book loan service, which results in a low visitor number in the library. Chinese children usually start reading at the age of 4, which is earlier than those in western countries. So many parents pay attention to their children's early reading and choose some random boring books for them. To deal with that, library stuffs should act as directors, and give suggestions for parents so as to get children interested and cultivate their correct habits for reading.

Finally, collaborative networks. Partners of Chinese libraries are different from those of American libraries. They cooperate with those agencies and organizations devoted in education for children such as local education bureau and the Communist Youth League. So they care much for the disadvantaged groups, which is another feature of Chinese libraries.[6]

4. Design practice for child-oriented community library

This project is located in a community in Tianjin, China. Directions for the design are set in terms of the analysis of age groups in the community. Based on the child-oriented service, children are divided into preschool children and school children.

The design should be based on different needs from different groups of readers. Besides designs of reading area for the elderly, accessible reading area, open reading area, semi-private reading area, discussion area for book clubs, family activity area, exchange area for child books, lecture hall, offices, front desk, warehouses and toilets...... detail design should also be attached great importance to.

4.1 Design principles

First, space. Too much inflexible segmentation of space may lead to psychological stress in children's mind. Based on the playful nature of children, some space should be designed as discussion areas, so that they won't disturb other readers and also give children a place to express themselves.

Second, color. Children have a strong ability to feel the change of colors, so the emotional effect of colors on children has to be concerned. It's better to choose some coordinated and vivid colors.

Third, environment. Children usually read for fun, so children's interest is a chief concern. Ergonomic facility for children is another important concern. Through obvious signs, lovely decorations and mini-sized facilities applicable to children, a welcoming atmosphere for reading will be created.

Fourth, safety. Just like banknotes, books in libraries that are frequently touched by readers may contain bacteria. Children are liable to diseases, so disinfection cabinets for regular sterilizations are necessary in libraries in order to prevent infectious diseases from books. Furthermore, children's libraries may experience peak periods of readers during weekends and holidays, so a restricted number of readers could be set because places with large crowds of people may easily to spread respiratory diseases. Preventive works are necessary.

Fifth, extended services. For communities without children's library, on-site services can be provided nearby especially in schools, residential areas and streets. Organizing various forms of activities is also a good idea. For example, through reading illustration books with their parents, children can learn about skills and methods in reading. Activities like that help strengthen the love between parents and kids, and they will have fun in reading.

4.2 Project analysis

Based on the preference of local residents, the source of inspiration comes from the shape of snail. There are two floors in the library. Floor one is for main activities in the library, while Floor Two works as the space dedicated to cultivate children's interest. (See Figure 1 and 2)

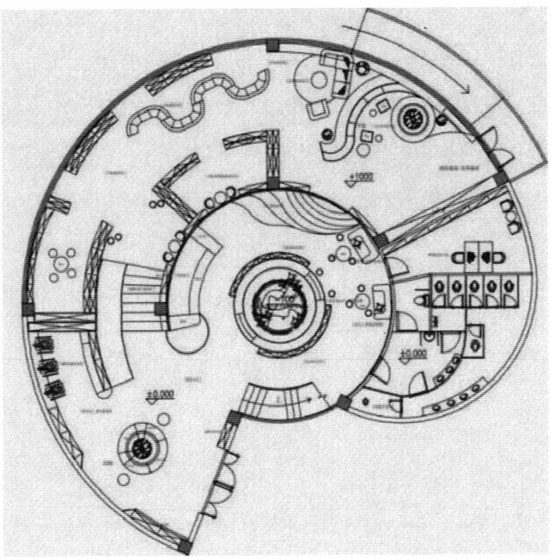

Figure 1. Plan of Floor 1

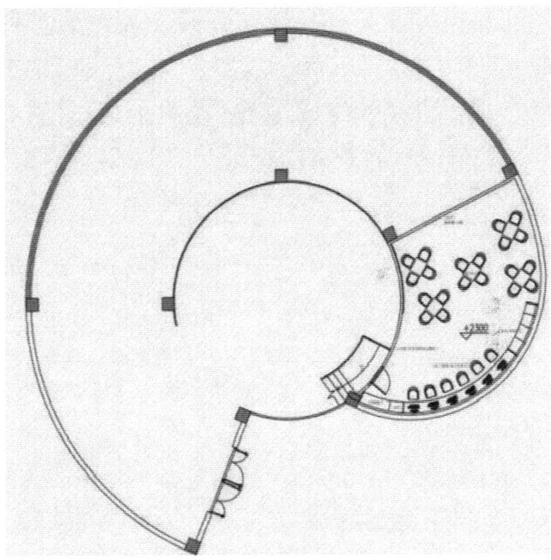

Figure 2. Plan of Floor 2

Preschool children are mostly between 2 to 5 years old. Forms of entertainment can be used to improve children's reading. The space are mainly designed as dynamic zones. (See Figure 3) The sunken space in the plan brings a sense of multi-levels, and also conforms to children's difficulty in concentration. Children may choose anywhere they like for reading thanks to the wooden floor with carpets on it, and family activities are also welcomed. Animal-shaped shelves work as an attraction to children, so that they will be interested in books, then love reading. According to children's preference on high color intensity, natural color is mainly used with some beige, yellow, orange and blue as decorations. Warm colors can greatly attract children's attention. (See Figure 4)

IOP Conf. Series: Materials Science and Engineering **520** (2019) 012022 doi:10.1088/1757-899X/520/1/012022

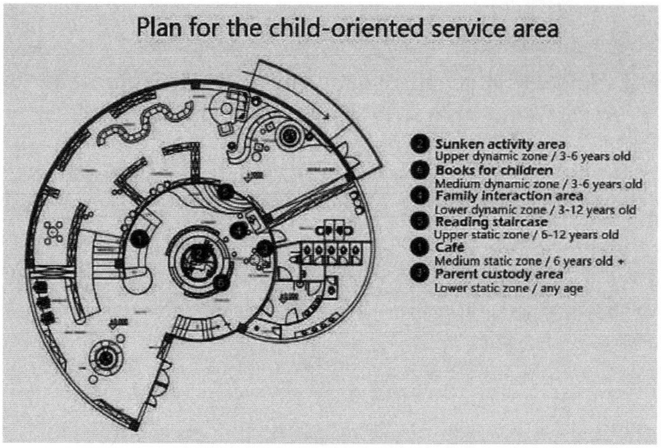

Figure 3. Plan for the children's reading area

Figure 4. Sunken Activity Area

The functions of school children area and preschool children area are quite different for their varied age groups and requirements for space. Obviously, two different kinds of reading area have been designed: a relatively open one and a relatively private one. (See Figure 5)

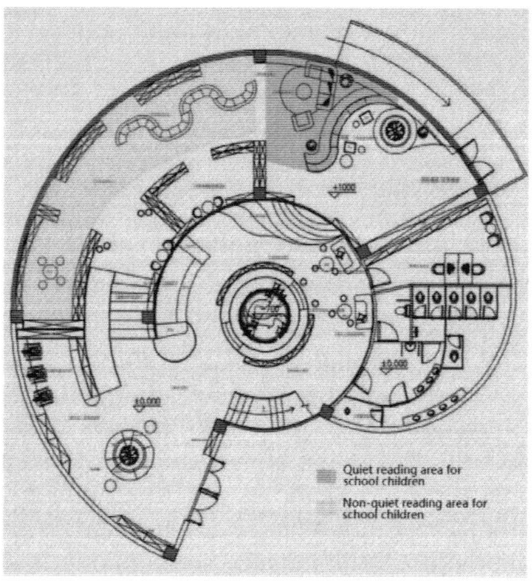

Figure 5. Quiet and non-quiet reading areas for school children

Figure 6. Open reading space for school children

Tall bookshelves made from acoustic materials separate the school children area from noisy preschool children, which is the solution to their requirement for individual privacy. They are able to learn, do their homework and discuss there. (See Figure 6)

5. Conclusion

For the safety of kids, streamlined design has been applied to the project. Fewer inflexible segments in the large space of library encourage the creativity of children. Diversified colors with high color intensity can attract children and make them interested. Sufficient lighting creates a bright environment for children to read. Ergonomic facilities in the library are fitted for children. Private space for independent learning and semi-private space for group discussion will satisfy both. Low saturation colors in school children area create a quiet environment for learning.

Besides the environment created through design, the effort of library staffs is also important for attracting children to read and cultivating their reading habits. Staffs should direct children to learn, and give them suggestions of right books to read. Community library should also regularly organize events like family activities, book reviews and promotions for latest books, to attract more children to start reading and love reading at an early age.

The service for children's library should focus on both readers and services, in order to provide child-oriented service and to satisfy children's requirements as much as possible.

References

[1] Z.hu dan,Community Library; *Overview, Evaluation and reflection.Library research*,12(2010)
[2] M.ei hong.*Individualized Service of American Community Library based on user demand orientation,Library research*,86(2015)
[3] W.u zhi nan,F.eng jia,*Public Library in Modern Public Culture Service system,Journal of Chinese Library*,41(2015)
[4] W.ang zi zhou,,*The rise and Development of Public Reading Space in China,Library and information knowledge*,02(2017)
[5] X.u huan,.*Study on the Establishment of Public Reading Space and the Development of Modern Library in China,Library construction*,01(2010)
[6] H.uang man li,.*Research on the Service Innovation of Children's Library readers,Science and Technology and economy in Inner Mongolia*,18(2014)